Other Titles in This Series

51 **Louis H. Kauffman, editor,** The interface of knots and physics (San Francisco, California, January 1995)

50 **Robert Calderbank, editor,** Different aspects of coding theory (San Francisco, California, January 1995)

49 **Robert L. Devaney, editor,** Complex dynamical systems: The mathematics behind the Mandlebrot and Julia sets (Cincinnati, Ohio, January 1994)

48 **Walter Gautschi, editor,** Mathematics of Computation 1943–1993: A half century of computational mathematics (Vancouver, British Columbia, August 1993)

47 **Ingrid Daubechies, editor,** Different perspectives on wavelets (San Antonio, Texas, January 1993)

46 **Stefan A. Burr, editor,** The unreasonable effectiveness of number theory (Orono, Maine, August 1991)

45 **De Witt L. Sumners, editor,** New scientific applications of geometry and topology (Baltimore, Maryland, January 1992)

44 **Béla Bollobás, editor,** Probabilistic combinatorics and its applications (San Francisco, California, January 1991)

43 **Richard K. Guy, editor,** Combinatorial games (Columbus, Ohio, August 1990)

42 **C. Pomerance, editor,** Cryptology and computational number theory (Boulder, Colorado, August 1989)

41 **R. W. Brockett, editor,** Robotics (Louisville, Kentucky, January 1990)

40 **Charles R. Johnson, editor,** Matrix theory and applications (Phoenix, Arizona, January 1989)

39 **Robert L. Devaney and Linda Keen, editors,** Chaos and fractals: The mathematics behind the computer graphics (Providence, Rhode Island, August 1988)

38 **Juris Hartmanis, editor,** Computational complexity theory (Atlanta, Georgia, January 1988)

37 **Henry J. Landau, editor,** Moments in mathematics (San Antonio, Texas, January 1987)

36 **Carl de Boor, editor,** Approximation theory (New Orleans, Louisiana, January 1986)

35 **Harry H. Panjer, editor,** Actuarial mathematics (Laramie, Wyoming, August 1985)

34 **Michael Anshel and William Gewirtz, editors,** Mathematics of information processing (Louisville, Kentucky, January 1984)

33 **H. Peyton Young, editor,** Fair allocation (Anaheim, California, January 1985)

32 **R. W. McKelvey, editor,** Environmental and natural resource mathematics (Eugene, Oregon, August 1984)

31 **B. Gopinath, editor,** Computer communications (Denver, Colorado, January 1983)

30 **Simon A. Levin, editor,** Population biology (Albany, New York, August 1983)

29 **R. A. DeMillo, G. I. Davida, D. P. Dobkin, M. A. Harrison, and R. J. Lipton,** Applied cryptology, cryptographic protocols, and computer security models (San Francisco, California, January 1981)

28 **R. Gnanadesikan, editor,** Statistical data analysis (Toronto, Ontario, August 1982)

27 **L. A. Shepp, editor,** Computed tomography (Cincinnati, Ohio, January 1982)

26 **S. A. Burr, editor,** The mathematics of networks (Pittsburgh, Pennsylvania, August 1981)

25 **S. I. Gass, editor,** Operations research: mathematics and models (Duluth, Minnesota, August 1979)

24 **W. F. Lucas, editor,** Game theory and its applications (Biloxi, Mississippi, January 1979)

23 **R. V. Hogg, editor,** Modern statistics: Methods and applications (San Antonio, Texas, January 1980)

22 **G. H. Golub and J. Oliger, editors,** Numerical analysis (Atlanta, Georgia, January 1978)

(Continued in the back of this publication)

AMS SHORT COURSE LECTURE NOTES

Introductory Survey Lectures

published as a subseries of

Proceedings of Symposia in Applied Mathematics

Proceedings of Symposia in
APPLIED MATHEMATICS

Volume 51

The Interface of Knots and Physics

American Mathematical Society
Short Course
January 2–3, 1995
San Francisco, California

Louis H. Kauffman
Editor

American Mathematical Society
Providence, Rhode Island

LECTURE NOTES PREPARED FOR THE
AMERICAN MATHEMATICAL SOCIETY SHORT COURSE
KNOTS AND PHYSICS

HELD IN SAN FRANCISCO, CALIFORNIA
JANUARY 2–3, 1995

The AMS Short Course Series is sponsored by the Society's Program Committee on National Meetings. The Series is under the direction of the Short Course Subcommittee of the Program Committee for National Meetings.

1991 *Mathematics Subject Classification*. Primary 18–XX, 53–XX, 55–XX, 57–XX; Secondary 78–XX, 81–XX, 82–XX, 83–XX.

Library of Congress Cataloging-in-Publication Data

The interface of knots and physics: an AMS short course on knots and physics / Louis H. Kauffman, editor.

p. cm. — (Proceedings of symposia in applied mathematics, ISSN 0160-7634; v. 51, AMS short course lecture notes)

Includes bibliographical references and index.

ISBN 0-8218-0380-8

1. Knot theory—Congresses. 2. Mathematical physics—Congresses. I. Kauffman, Louis H., 1945– . II. American Mathematical Society. III. Series: Proceedings of symposia in applied mathematics; v. 51. IV. Series: Proceedings of symposia in applied mathematics. AMS short course lecture notes.

QC20.7.K56I58 1995
530.1′54224—dc20

95-37130
CIP

∞ The paper used in this book is acid-free and falls within the guidelines
established to ensure permanence and durability.
♻ Printed on recycled paper.

10 9 8 7 6 5 4 3 2 1   01 00 99 98 97 96

Contents

Preface

This book is the result of an AMS "Short Course on Knots and Physics" that was held in San Francisco on January 2 and 3, 1995, just prior to the meeting of the American Mathematical Society. As the reader will quickly see, the range of this course goes well beyond knots to the study of invariants of low dimensional manifolds, and even to hints about possible extensions of this work to higher dimensional manifolds. The pervasive theme is the use of ideas and methods of mathematical physics to extract topological information about knots and manifolds.

Here is a quick description of the contents of the book.

The first paper, "Knots and Statistical Mechanics" by Louis H. Kauffman is a review of basic knot theory and its relationships with discrete statistical mechanics models. These relationships give sharp motivations for the construction of the original Jones polynomial and its generalizations. In fact, as the paper shows, it is possible to take the idea of a state summation with the compass of topological invariance and rediscover interesting technicalities such as specific solutions to the Yang-Baxter equation and motivated beginnings of the theory of quantum groups. This paper provides an entry point into the subject from combinatorial topology. The paper ends with a comparison of piecewise-linear and oriented amplitude models. It simplifies previous formulations of quantum link invariants derived from solutions to the Yang-Baxter equation.

The second paper, "An Introduction to Topological Quantum Field Theory" by Ruth J. Lawrence, sketches the history of topological quantum field theories from their inception in Edward Witten's quantum field theoretic interpretation of the Jones polynomial and his concurrent generalization to invariants of three dimensional manifolds. Witten's original definitions in terms of functional integrals were axiomatized by Atiyah and Witten to give an abstract definition. Such theories were shown to exist on purely combinatorial grounds by Reshetikhin and Turaev using quantum groups, by Khono and Crane by using rational conformal field theory (following Witten's own reformulation of the functional integral), by Turaev and Viro using quantum recoupling theory. These combinatorial definitions partake of a mix of algebra, categories and generalized state summations. This paper gives a clear exposition of the range of ideas with many specific examples, including formulations of the way ahead for four dimensional manifolds and beyond.

The third paper, "Vassiliev and Quantum Invariants of Braids" by Dror Bar-Natan, proves that braid invariants coming from quantum $gl(N)$ separate braids by recalling that these invariants (properly decomposed) are all Vassiliev invariants, showing that all Vassiliev invariants of braids arise in this way, and reproving that Vassiliev invariants separate braids. The paper contains a very concise introduction to Vassiliev invariants and an excellent bibliography for the reader interested in further information on this topic. Vassiliev invariants are a fundamental way to construct knot invariants. They can be used as building blocks for all the known quantum link invariants. At the present time it is not known whether Vassiliev invariants yield a larger class of invariants than those derived from solutions to the Yang-Baxter equation (as described in the first paper in this volume).

The fourth paper, "The Modern Legacies of Thomson's Atomic Vortex Theory in Classical Electrodynamics" by Samuel J. Lomonaco, is a survey of the interaction of knot theory and electromagnetic theory from Lord Kelvin's knotted vortices to modern knotted electrostatic vortices and the problems of minimal knotted tubes of electrical charge. It is tantalizing to compare the knots and physics of this paper with the knots and physics in the previous papers. In considering knotted tubes of electrical charge one poses a problem in classical physics that involves knotting. It is a problem that makes mathematical knots more physical by adding extra structure to them. In considering state summations and topological quantum field theories, the knots are not made more physical but it is seen that deep physical ideas apply to their topological structure. It is an open problem to find the significant relations between these two approaches to the physicality of knots.

The fifth paper "Spin Networks in Nonperturbative Quantum Gravity" by John Baez, is a survey of how knots and generalized spin networks have come to play an important role in certain approaches to the unification of general relativity and quantum physics. In particular the paper describes the remarkable theory due to Ashtekar, Smolin and Rovelli where test particles moving around knots measure significant data about the quantum states of gravity. The use of spin networks is a modern evocation of an idea of Roger Penrose from 35 years ago. The Penrose spin nets were a combinatorial version of the recoupling theory of $SU(2)$. They naturally generalize to recoupling theory for quantum groups and play a direct role in the topological quantum field theories associated with the Jones polynomial. In this paper, the spin networks are put to use in the service of states and geometry of quantum gravity.

I would like to take this opportunity to thank all the speakers in the Short Course for their very dedicated work in teaching the course and preparing the papers for this volume. A big round of applause goes as well to all the participants who were willing to listen, ask questions, and keep us going in this exposition of complex terrain.

Louis H. Kauffman
May 1995

Proceedings of Symposia in Applied Mathematics
Volume **51**, 1996

KNOTS AND STATISTICAL MECHANICS

by Louis H. Kauffman

Abstract. This paper discusses the role of state summations in the construction of invariants of knots and links. Such state summations arise in statistical mechanics as partition functions and as quantum amplitudes. The Jones polynomial and its generalizations are intimately related to these techniques from mathematical physics.

1. Introduction

This essay is an introduction to relations between knot theory and statistical mechanics with emphasis on the construction of invariants of knots and links. We show how state sums and solutions of the Yang-Baxter equation can be used to construct the Jones polynomial and two of its generalizations, the Homfly and Kauffman polynomials. We emphasize in this exposition that in these cases the relevant solutions to the Yang-Baxter equation arise quite naturally in the course of considering skein relations for the invariants. This circumstance is not the case in general, but it provides a convenient way to enter the subject from the side of the topology. Aside from a short introduction to quantum groups in section 6, these notes do not discuss relations between quantum groups and link invariants.

The paper is organized as follows: Section 2 reviews background on the combinatorial basis for knot theory. In particular , it provides a reference for those combinations of (Reidemeister) moves that generate ambient isotopy and regular isotopy. Section 3 gives background on the two two-variable skein polynomials that generalize the Jones polynomial. Section 4 discusses a first state model - the bracket polynomial [34]. The bracket gives an elementary construction for the original Jones polynomial, and it provides a pattern for more complex state models. The bracket is constructed without using the notion of a Yang-Baxter equation, but once we have it then a number of solutions to the Yang-Baxter equation appear naturally. This section discusses the relationship between the bracket expansion and the Potts model in statistical mechanics.

Section 5 introduces the context that I call *abstract tensors.* A tensor has indices -- an abstract tensor has lines attached to its body. Tensors are contracted by matching their indices and summing over the matched values -- abstract tensors are contracted by connecting their lines. Abstract tensors are an intermediate category between knot theory and state models. A link diagram can be given the structure of an abstract tensor. Section 6 gives a model for the bracket as a "vacuum-vacuum expectation" and in the process motivates a special solution to the Yang-Baxter equation that comes from the quantum group associated with SL(2). This solution to the Yang-Baxter equation is generalized and utilized in the next two sections.

Sections 7 and 8 give specific state models related to the two skein polynomials and to some generalizations. Section 8 builds upon section 7, with the Yang-Baxter solution for SO(n) emerging from an extension for the model already obtained in section 7. In this case the state model is simpler than the tensor that emerges from it! An extra complexity in this tensor (for SO(n)) is explained geometrically in terms of a compensation for rotation number.

Section 9 then considers the geometrical structure of these models,showing how they can be constructed via strictly local contributions in the piecewise linear setting. In this context, the rotation compensation, alluded to above, appears locally as an angle compensation.

Section 10 is a condensed exposition of the structure of oriented state sums as amplitudes in the sense of section 6. In this section we show how to rewrite the piecewise linear models of section 9 as oriented amplitudes in the spirit of section 6. This produces a neat simplification and an explanation of the genesis of the spin preservation conditions for the piecewise linear models.

This paper is a revised version of [44]. More information along the lines of this paper can be found in [44], [39] and many of the references in the index. The subject is by now quite vast, and I must offer an apology to many researchers whose names certainly should appear in the index and do not appear due to its incomplete nature. It should be mentioned that the models in sections 7 and 8 are originally due to Vaughan Jones in a somewhat less graphical form and that the piecewise linear models in section 9 are originally due to Vaughan Jones in a continuous form. The piecewise linear version is due to the author of this paper and appeared for the first time in [44]. The point of view taken in section 10 is a simplification of previous work in this field.

The paper does not have to be read in strictly sequential order. For the reader interested in first steps, I recommend sections 1 through 4. For the reader interested in the abstract structure of the models, I recommend sections 4,5,6 and 10. For the reader already interested in the skein polynomials and curious about how they are modelled by state sums the sections 7,8 and 9 will be particularly useful. Many readers may want to go to section 10 first, since it describes a direct line into these models.

It gives the author great pleasure to thank Claude Bourgeois for a key conversation about the relationship of topological amplitudes and piecewise linear state sums. This relationship is explained in section 10 of the present paper.

2. Link Diagrams and Regular Isotopy

A *link* is an embedding of a collection of circles into the oriented three-dimensional sphere or, equivalently, into three-dimensional Euclidean space. A link of n circles is said to have n components, and a 1-component link is called a *knot* if it cannot be continuously deformed into a standard (unknotted) circle (represented as a unit circle in one of the coordinate planes in Euclidean three-space. One refers to this standard unknotted circle as *the unknot*, and to a link consisting of a collection of separated unknots as a *trivial link* or an *unlink of n components.*

The notion of continuous deformation for links in three-space is called *ambient isotopy.* Two links are said to be *ambient isotopic* if there is a family of embeddings, parametrized by t in the interval of real numbers from zero to one, such that the embedding at zero is the first link and the embedding at one is the second link. These embeddings must behave continuously as a function of the parameter t; in other words, the corresponding mapping from the cross product of the collection of circles and the unit interval to the three-space is required to be continuous.

It is assumed here that all knots and links are *tame.* This means that they can be represented by differentiable curves in space or, equivalently, by a piecewise linear embedding composed of straight line segments in space.

The theory for tame links can be elegantly reformulated in terms of diagrams for the links projected into a plane. A *link diagram* (See Figure 1) is a drawing of a projection of the link with at most two points of different segments mapped to the same point. These segments have distinct tangent vectors, so that they cross one another at the point of intersection. At such a crossing, one of the segments is drawn with a small break to indicate that it *crosses under* the other one. (See Figure 1)

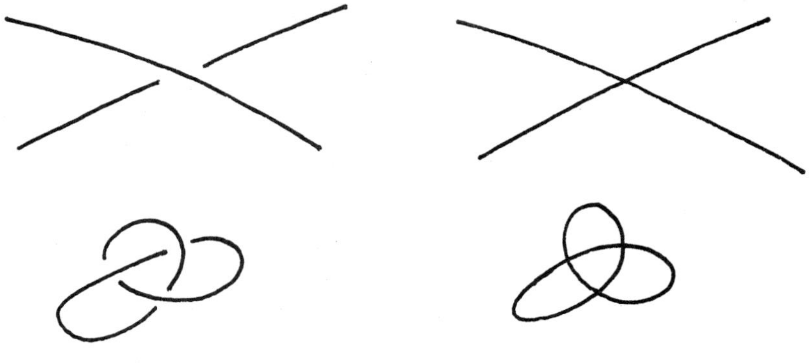

Figure 1

Such a diagram can be regarded as a locally 4-valent plane graph with extra structure at the vertices (indicated by the under/over crossing convention). I shall refer to a (locally 4-valent) plane graph *without* this extra structure as a *universe.* Thus Figure 1 illustrates a trefoil knot and its associated universe. A universe with n vertices can be configured as a link diagram in 2-raised to the nth power ways by making the appropriate 2-fold choice (over or under) at each crossing. Many of these choices are unknotted and some may be equivalent to one another.

The knot or link K^* obtained from a given diagram K by switching all the crossings is called the *mirror image* of K. (The mirror is the plane itself!) A link that is ambient isotopic to its mirror image is said to be *achiral.* If it is not ambient isotopic to its mirror image, then K is *chiral.* The trefoil knot shown in Figure 1 is chiral. Generally, chirality is a subtle property to determine, and the discrimination of links up to ambient isotopy is an incomplete subject as of this writing.

This brings us to the diagrammatic reformulation of ambient isotopy due to Reidemeister ([58], [10]). Reidemeister showed that the three elementary diagram moves (the *Reidemeister moves*) shown in Figure 2, suffice to describe ambient isotopy. That is , Reidemeister proved the

Theorem 2.1. *Two diagrams represent ambient isotopic links if and only if one diagram can be transformed to the other by a finite sequence of Reidemeister moves.* (See Figure 2.)

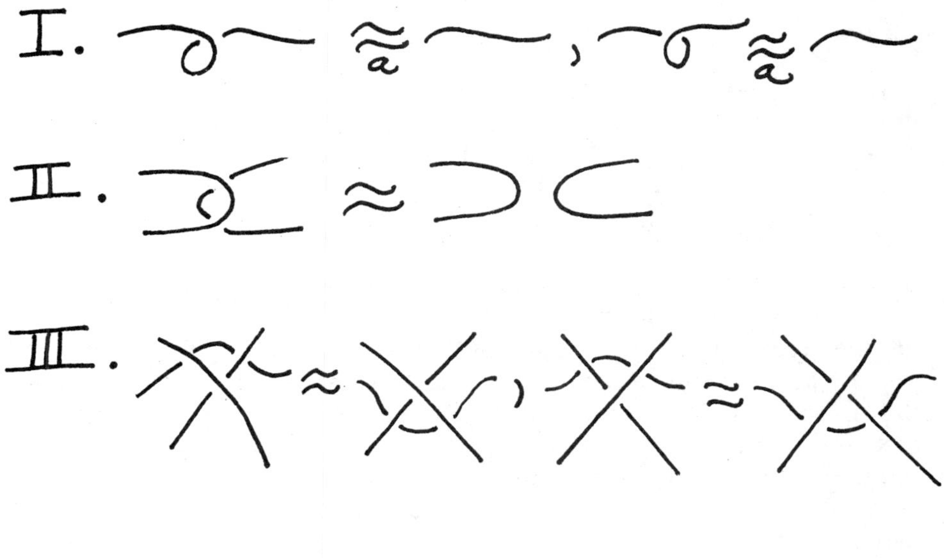

The Reidemeister Moves
Figure 2

Remark. Some remarks are needed to interpret this theorem. First the moves illustrated in Figure 2 are to be performed so that they only affect the small part of the diagram that is indicated. Each small diagram in the Figure represents a part of a larger diagram, and diagrams connected by a sign of equivalence ($\approx$) differ only in the small diagrams shown in the figure-- the outsides are identical. No extra lines are allowed in the domains of the moves. Thus the deformation shown below requires more than one move, even though it can be regarded as a "higher order" type I move.

Finally, we allow topological deformations of the underlying graph that do not change the crossing structure of the diagram. These deformations could be termed "move zero" and indicated via

Later, we shall consider the piecewise linear versions of the moves. Then move zero can be directly articulated. Here it is convenient to put move zero in the background by asking the reader to put on the glasses of a topologist who cannot tell the difference between a Jordan Curve and a circle.

Simplification of the Oriented Moves

In the case of oriented links we must discriminate two versions of the type II move , and two basic versions of the type three move (along with a number of different choices for crossings at the triangle of the three-move. In Figure 3 I have indicated the *oriented* type II move as IIA and the *reverse oriented* type II move as IIB. The figure illustrates two type three moves labelled IIIA, and IIIB respectively. In the move IIIA the inner triangle orientations do not form a cycle, and the move will be called the *non-cyclic* type III move. In the move IIIB, the inner triangle is oriented in a cycle, and the move is called a *cyclic* type three move.

Crossings in an oriented diagram are called *positive* or *negative* according to the convention shown in Figure 3. (A crossing is positive if the overcrossing segment coincides in orientation with the undercrossing segment when the overcrossing segment is turned counterclockwise.)

In the Figure 3, the moves IIIA and IIIB are shown with all positive crossings. Call these *positive* moves of type IIIA and IIIB. Call them of *negative* type if all the crossings are negative.

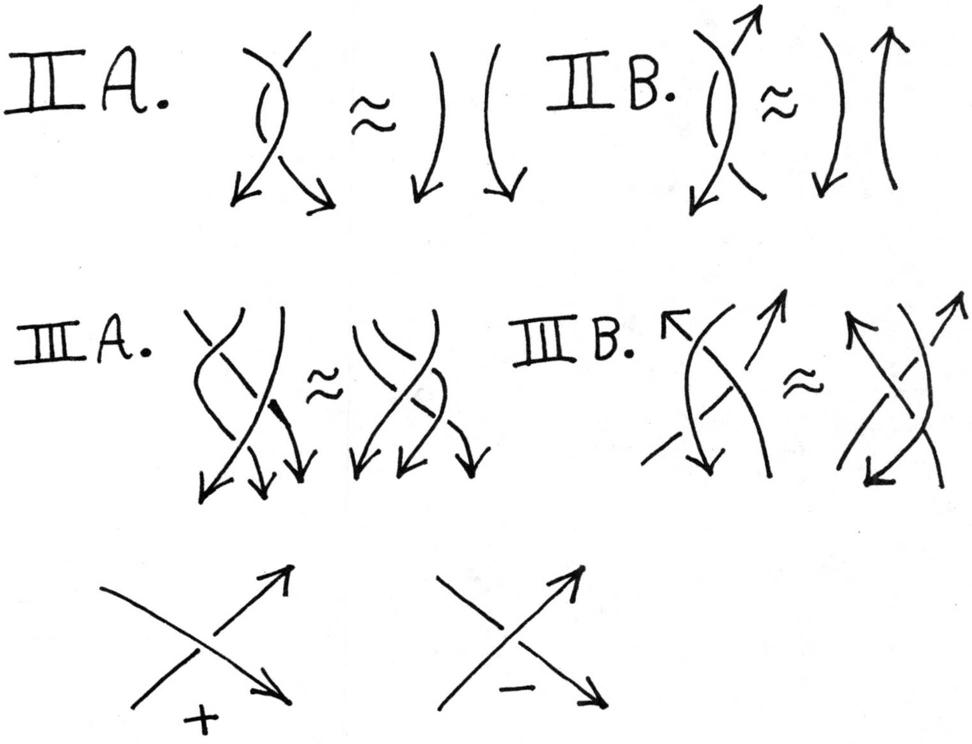

Oriented Reidemeister Moves
Figure 3

Lemma 2.2. *In the presence of the moves of type IIA and IIB it is sufficient to have only the moves IIIA and IIIB of positive and negative type. That is , the other type three moves can be generated from the moves IIA,IIB and IIIA IIIB (positive and negative).*

Proof. View Figure 4.

Lemma 2.3. *In the presence of moves IIA and IIB, the move IIIB (positive,negative) can be generated from IIIA (positive, negative).*

Proof. View Figure 4.

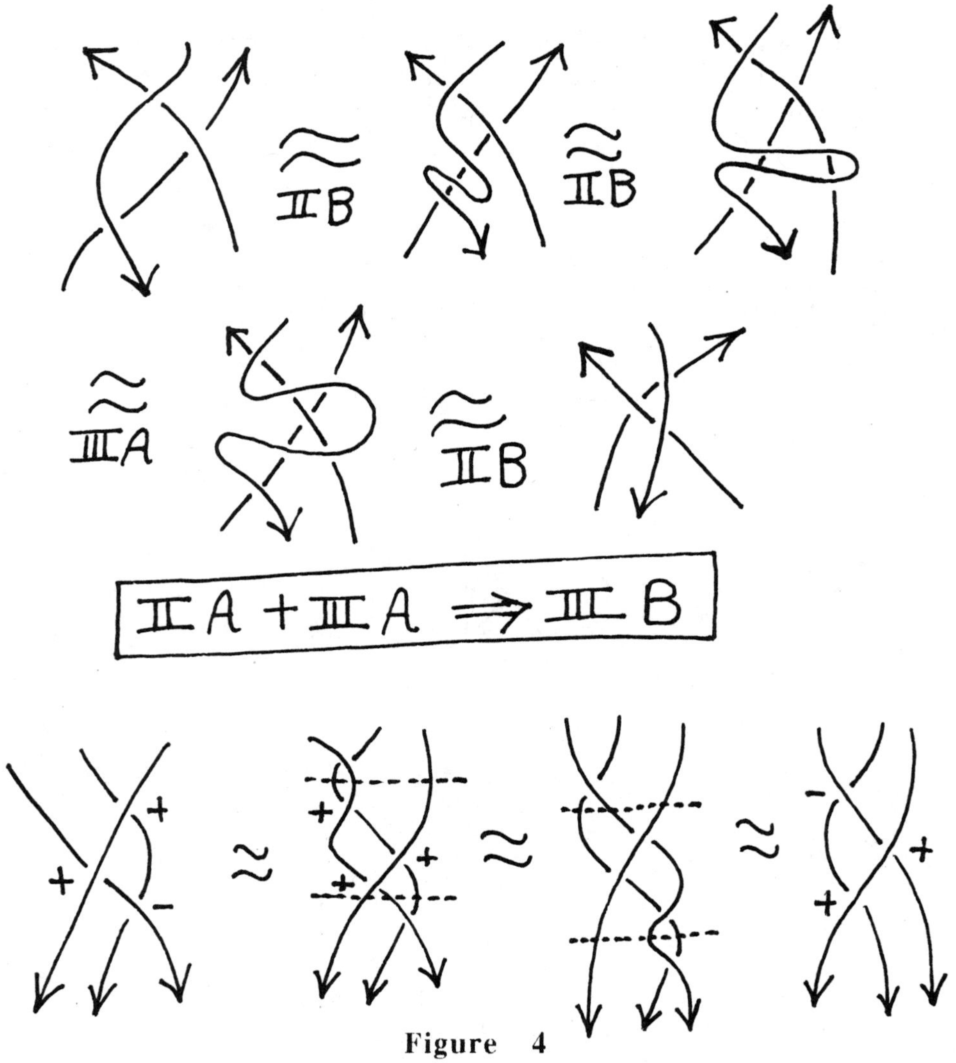

Figure 4

Thus we have proved the

Theorem 2.1. *The moves I, IIA, IIB, and IIIA (positive, negative) form a complete set of generating moves for the ambient isotopy of oriented link diagrams.*

Regular Isotopy

A further reduction is available. With suitable restrictions, we can also get rid of the type I move. Call *regular isotopy* the equivalence relation on link diagrams that is generated by the moves of type II and type III only.

A regular isotopy appears as a motion of the link diagram that is almost entirely planar. In fact, a regular isotopy projects to a *regular_homotopy* of the underlying universe of the diagram ([38],[68]). As a result, one simple invariant of regular isotopy is the Whitney degree [68] or *rotation number* of the underlying plane curve. The Whitney degree of K, here denoted *rot(K)*, counts algebraically the number of turns of the unit tangent vector to the curve. In terms of diagrams the rotation number can be computed by the algorithm:

1. Form the collection of *Seifert circuits* associated to K by *smoothing* each crossing as shown below.

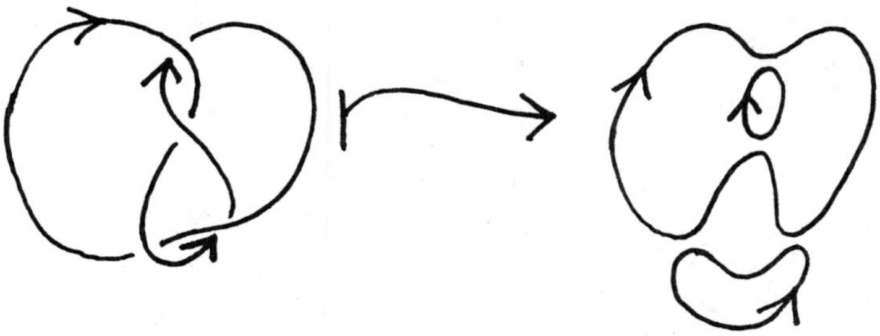

Let S(K) denote the collection of oriented circuits in the plane obtained by this operation.

2. Define rotation number for simple closed curves as indicated below (positive is counterclockwise, negative is clockwise).

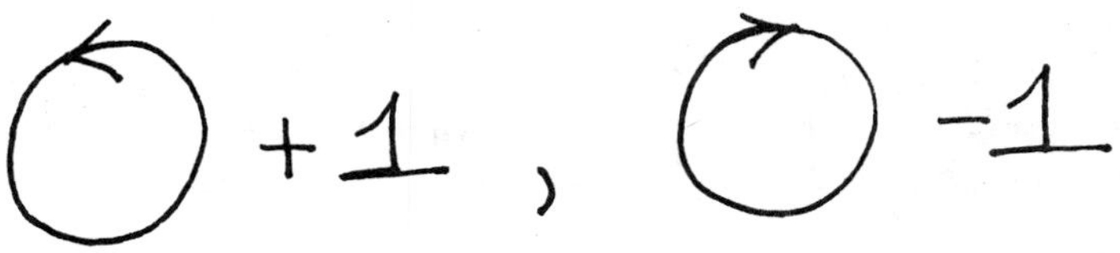

3. Define rot(K) for an oriented link diagram to be *the sum of the rotation numbers of the circuits in S(K).*

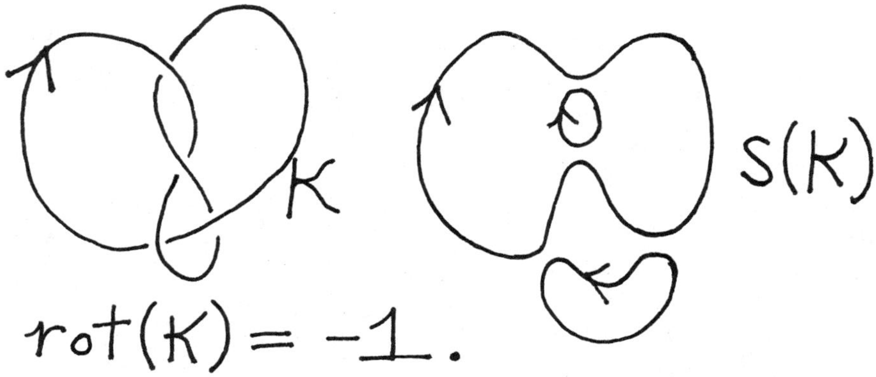

It is easy to see that this definition coincides with the original definition of Whitney in the case of immersed plane curves. It is also easy to verify that rot(K) is an invariant of regular isotopy.

The rotation number does not depend upon the crossing structure in the link diagram. The next invariant of regular isotopy, the *writhe*, w(K), adds up the signs of the crossings:

Definition 2.4. *The writhe , w(K), of an oriented link diagram K is the sum of the signs of the crossings of K.*

The writhe is obviously invariant under the II and III moves, hence it is an invariant of regular isotopy.

Example.

The following theorem (compare [66]) is the primary reason for introducing writhe and rotation number in the context of regular isotopy.

Theorem 2.5. *Let K and K' be oriented knot diagrams with the same writhe and the same rotation number (i.e. w(K)=w(K') and rot(K)=rot(K')). Suppose that K is ambient isotopic to K'. Then K is regularly isotopic to K'.*

Theorem 2.6. *Let K and K' be link diagrams such that every component of K or of K' has writhe zero and rotation number one. Suppose that K and K' are ambient isotopic links. Then K and K' are regularly isotopic diagrams.*

Proposition 2.7. *For any knot diagram K, the difference between the writhe and the rotation number is odd. That is, rot(K) = w(K) + 1 (modulo 2). Therefore it is possible to adjust the writhe of K by taking its connected sum with a diagram U that is unknotted (U can be a sum of simple curls) so that rot(K#U) = 1 and w(K#U) = 0.*

Remark. If we want to determine whether two links L and L' are ambient isotopic, it is sufficient to adjust diagrams of L and L' (without disturbing their ambient isotopy class) so that every component has writhe zero and rotation number one. So adjusted, if L and L' are ambient isotopic links then their diagrams will be regularly isotopic. Since Proposition 2.7 tells us that these adjustments can be made, this shows that, with such adjustments, the type I move can be eliminated from the theory.

Discussion. Before proving these results, it is convenient to discuss some properties of knot diagrams that figure in these proofs and in our later discussions.

1. Any knot can be turned into an unknot by switching a finite number of its crossings.

The proof is seen by reference to a *standard unknot*, U, obtained from the underlying universe of the knot K by traversing it - always overcrossing on the first pass through any crossing. See Figure 5. The standard unknot is unknotted, and it can be obtained from K by switching the subset of crossings where it differs from K.

2. Any link can be unlinked by a finite number of crossing switches. To unhook two components, switch mutual crossings so that one component lies entirely over the other one.

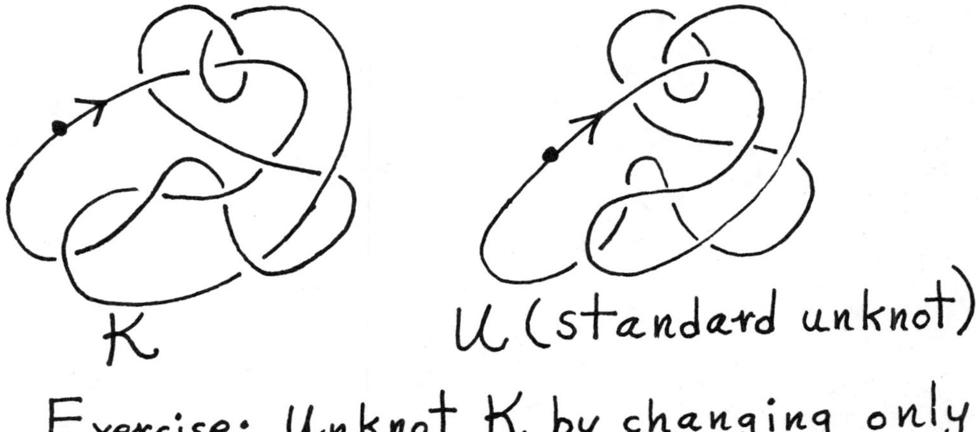

Exercise: Unknot K by changing only one crossing.

Figure 5

3. Let *regular homotopy* be the equivalence relation on universes (link diagrams without over and under crossing designations) generated by the shadows II' and III' of the type II and type III moves. See Figure 6.

Theorem. *Any knot universe is regularly homotopic to a connected sum of curl shadows.*

Proof. Let U be the universe and U' a standard unknot overlying U. Then U' is regularly isotopic to a connected sum of curls (exercise!). Take the shadow of this regular isotopy. This completes the proof.

4. As discussed previously, the set of Seifert circuits for an oriented universe U, is obtained by splitting all the crossings of U, and taking the resulting set of Jordan curves in the plane. Given a universe U, let s(U) denote the number of Seifert circuits for U.

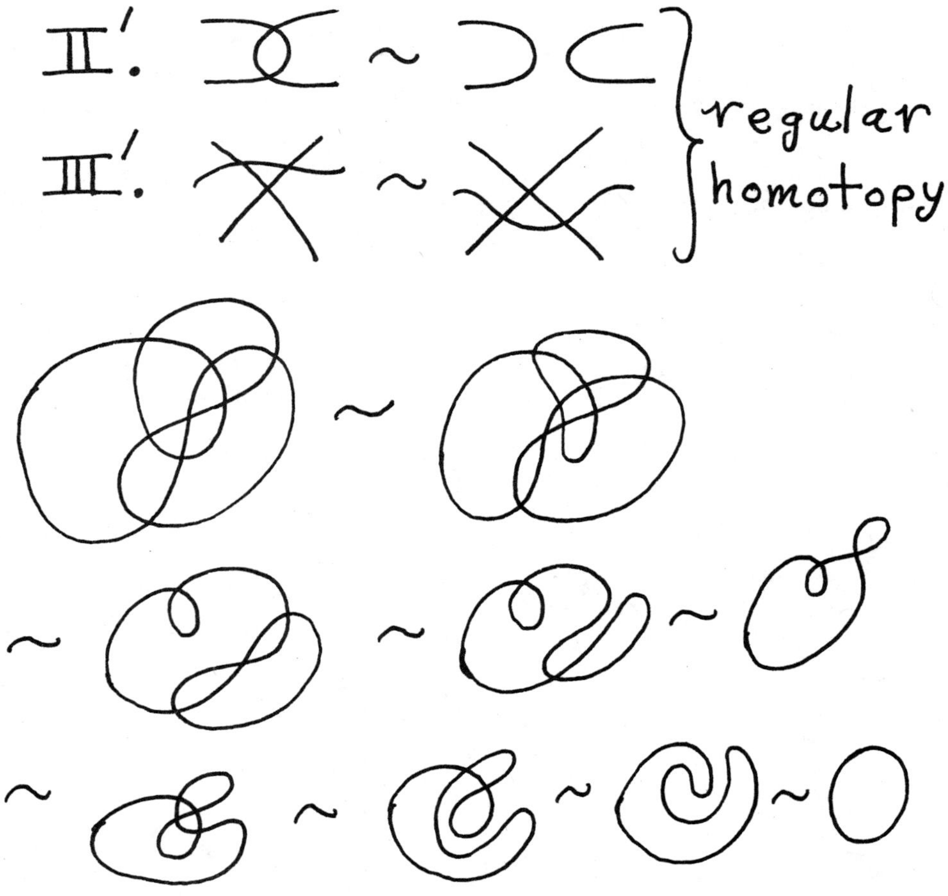

Figure 6

Proposition. *If universes U and V are regularly homotopic, then s(U) and s(V) are congruent modulo 2. The parity of the number of Seifert circuits is invariant under regular isotopy.*

Proof. Check that this is so for each of the moves II' and III'.

5. The parity of the number of vertices of a universe is invariant under regular homotopy.

6. Proposition. *If v(U) denotes the number of vertices of a universe U, and s(U) denotes the number of Seifert circuits of U, then, if U is a shadow of a knot diagram (a "one-component universe") , then s(U)-v(U) = 1 (modulo 2).*

Proof. By 4. and 5. we may assume that U is a connected sum of curl shadows. The result follows at once.

Theorem. *Let K be a knot diagram, then rot(K) - w(K) = 1 (modulo 2).*

Proof. Let U be the underlying universe for K. Then rot(K) = s(U) (modulo 2) and w(K) = v(U) (modulo 2). Apply ~~5.~~ 6

7. View Figure 7. This figure shows that , by adding a curl, one can raise or lower the writhe and rotation number of a link component independently of each other. This information, together with ~~6.~~ completes the proof of Proposition 2.7.

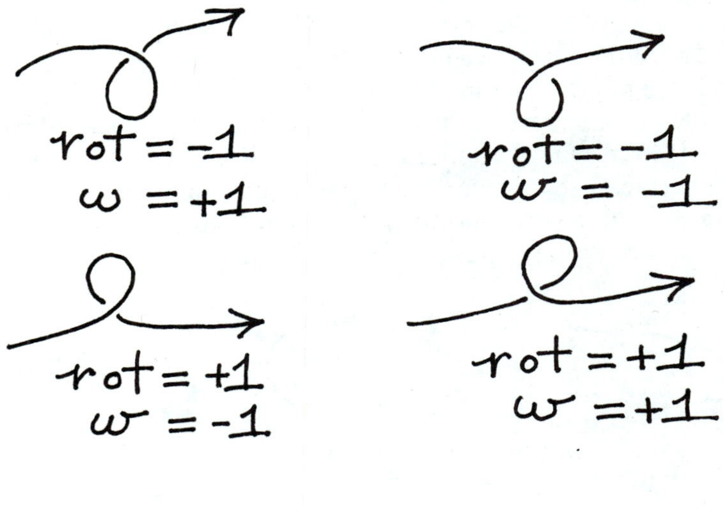

Figure 7

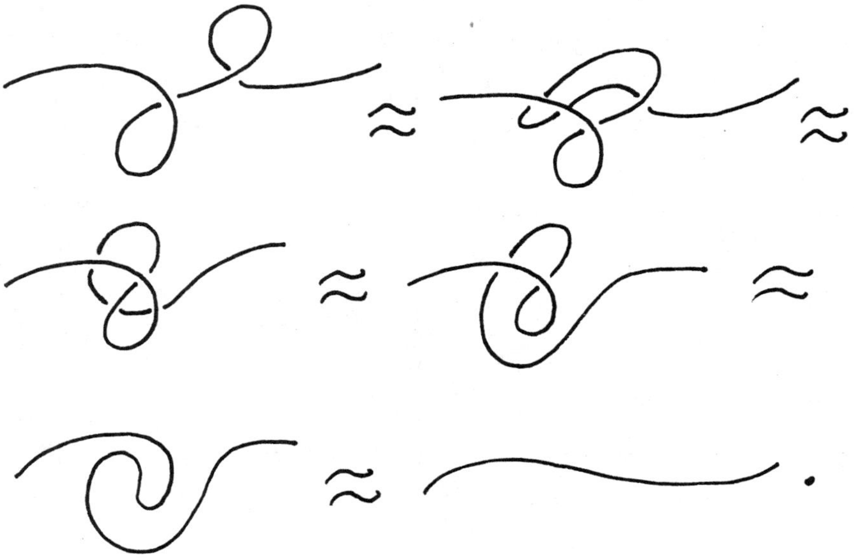

Generalized Whitney Trick
Figure 8

It remains to prove Theorems 2.5 and 2.6.

Proof of 2.5. Note that curls of opposite rotation number and writhe cancel each other as shown in Figure 8. This is a generalization of the Whitney trick [68].

In the first part of the proof I shall assume that K and K' are both unknotted with w(K)=w(K') and rot(K)=rot(K'). We must show that K is regularly isotopic to K'.

Since K is unknotted, there is a sequence of Reidemeister moves taking it to a simple closed curve. Since type I moves are strictly local, we can conserve them until the end of the isotopy, at the cost of adding a few more type II and type III moves. This may involve sliding an arc past a curl as illustrated below:

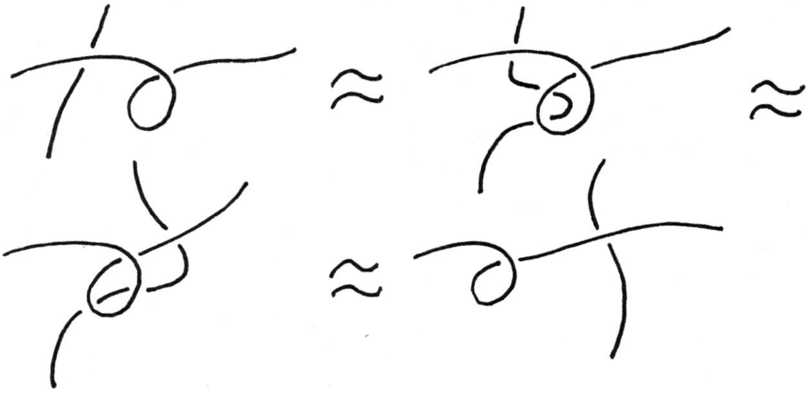

By using this technique, any unknot diagram is regularly isotopic to a diagram consisting entirely of curls. Note that curls can be permuted past one another:

Thus we can bring together curls that are susceptible to the influence of the the generalized Whitney trick, cancelling them in pairs. Then there are two cases:

1. rot < w

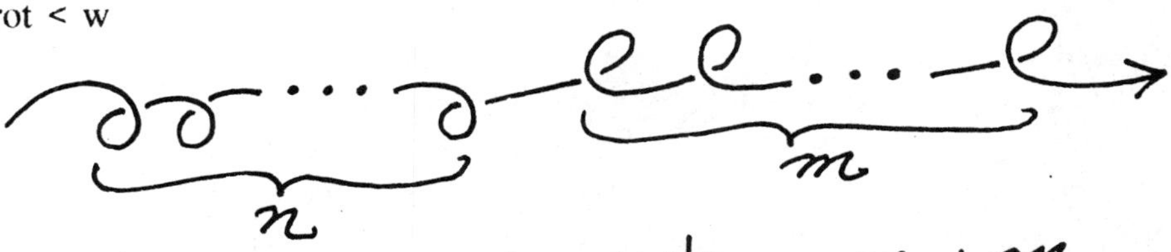

$$w = n + m, \quad rot = -n + m.$$

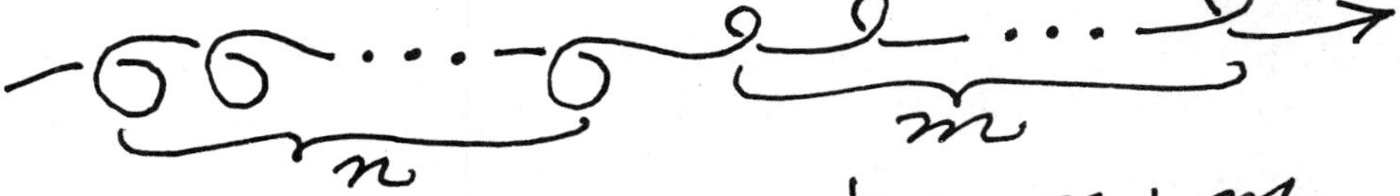

$$w = n - m, \quad rot = -n + m.$$

2. rot >= w

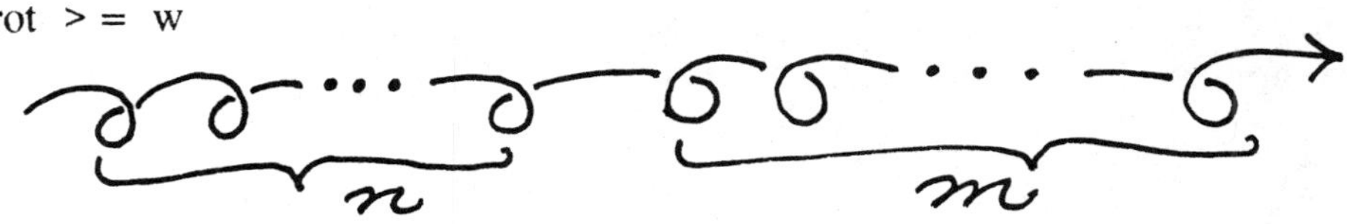

$$w = n - m, \quad rot = -n - m.$$

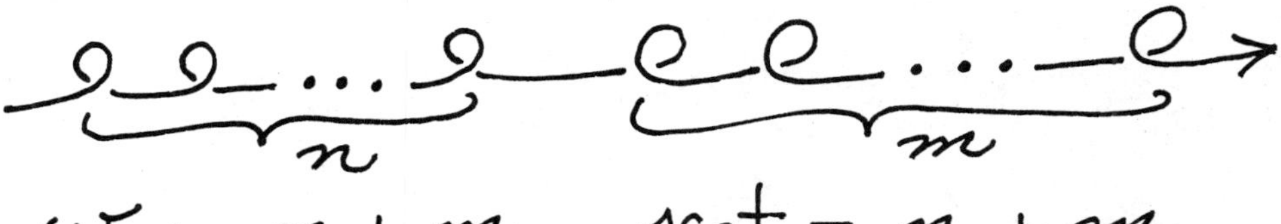

$$w = -n + m, \quad rot = n + m.$$

These give a set of canonical forms, from which the result for unknots follows.

Now suppose that K and K' are any two knot diagrams with the same writhe and rotation numbers. If K is ambient isotopic to K', then there is a diagram K" so that

$$K \approx K'' \# U$$
$$K' \approx K'' \# V$$

($\approx$ denotes regular isotopy.)
where U and V are unknot diagrams (connected sums of curls will suffice). This follows via conservation of the type I move. It then follows from our hypotheses that U and V have the same writhe and the same rotation number. Therefore , U and V are regularly isotopic. Hence K and K' are regularly isotopic. This completes the proof of Theorem 2.5.

Proof of 2.6. The proof follows the same pattern as the last part of the argument for 2.5. The ambient isotopy of the two links can be factored into a regular isotopy followed by a collection of type I moves. This means that if K has components $K_1, K_2, \ldots K_n$ then the components of K' can be labelled $K'_1, K'_2, \ldots K'_n$ so that there are unknots $U_1, \ldots U_n$ and $V_1, \ldots V_n$ with K_i regularly isotopic to $K''_i \# U_i$ and K'_i regularly isotopic to $K''_i \# V_i$ for $i=1,\ldots,n$. That is , the link K is regularly isotopic to the link obtained from a union of components $K''_i \# U_i$ and K' is regularly isotopic to the link with components $K''_i \# V_i$. The hypotheses on the writhe and rotation number then show that U_i and V_i have the same writhe and rotation number for each i. This implies that U_i and V_i are regularly isotopic, and hence that K and K' are regularly isotopic, because these individual isotopies can all be assembled into one large regular isotopy between the two links. This completes the proof of Theorem 2.6.

3. The Two-Variable Skein Polynomials

This section reviews the properties of two two-variable skein polynomials - the Homfly polynomial ([17],[23],[50]) and the Kauffman polynomial [38]. Each of these polynomials generalizes the original Jones polynomial [28]. The Homfly polynomial is also a generalization of the Alexander-Conway polynomial. The Kauffman polynomial generalizes the one-variable unoriented polynomial of Millett, Brandt, Lickorish [9] and Ho [22].

The Homfly polynomial, P_K, is described via the equation

$$P_K = a^{-w(K)} H_K$$

where the polynomial H_K is an invariant of regular isotopy satisfying the equations

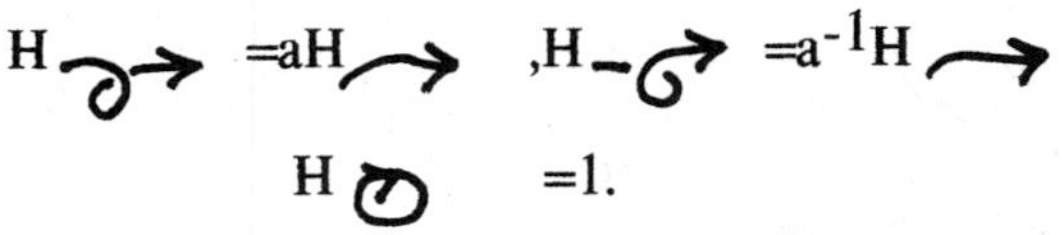

For a=1, the polynomial H_K coincides with the Alexander-Conway polynomial. The description of the Alexander polynomial in the form of an exchange identity is due to Conway [12]. The Conway version normalizes the Alexander polynomial, and extracts more information from it in the process (It can detect chirality for some links with an even number of components.).

The Homfly polynomial, an ambient isotopy invariant, can be expressed via the direct identity

(*) 

The original Jones polynomial, $V_K(t)$, corresponds to the substitution in the Homfly polynomial $a=t$, $z = \sqrt{t} - 1/\sqrt{t}$:

Remark on Framing

It is of interest to interpret the extra variable a in the Homfly polynomial as measuring framing for the link. This framing information can be regarded as an interpretation of the writhe , w(K). That is , replace the link diagram by thickening each link component into a "flat" band.

This flat band is obtained by drawing a second curve parallel to the original curve for each component and regarding the two parallel curves as forming an embedded band in three-space. For a knot, the writhe of the diagram is equal to the linking number of the two edges of the flat band. For a link, the writhe contributions between components of the link give twice the linking number of these components. Thus we have the formula

$$w(K) = \sum \text{lk(band boundaries)} + 2\sum \text{lk(band to band)}.$$

From the point of view of the bands, w(K) is invariant topological information about the framing. (A framing of the original link is equivalent to the specification of a band embedding.)

The appearances of the variable a in the skein identity of the Homfly polynomial (*) can be regarded as compensations so that the total framing is set to zero for each member of the skein triple.

It follows from the rules for these polynomials that

$$H_{0\,K} = \delta\, H_K, \qquad \delta = (a-a^{-1})/z$$

and similarly for P.

The Homfly polynomial has many properties. See [17] , [23] and [50] for more information.

The Kauffman polynomial

$$F_K[a,z]$$

is also defined as a writhe-normalized version of a regular isotopy invariant polynomial.

$$F_K = a^{-w(K)} L_K$$

Here the polynomial

$$L_K(a,z)$$

is a regular isotopy invariant for unoriented links , and satisfies the unoriented skein identities

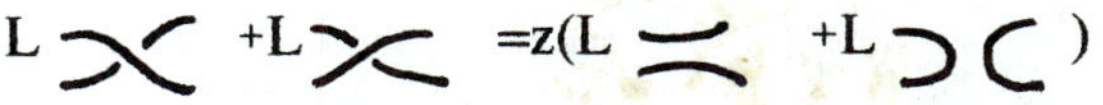

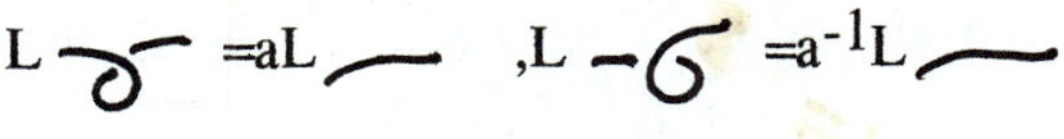

In this paper we shall use an equivalent version of the F and L polynomials (see [38]) that I call the Dubrovnik polynomial. This polynomial D is a regular isotopy invariant for unoriented links satisfying the identities:

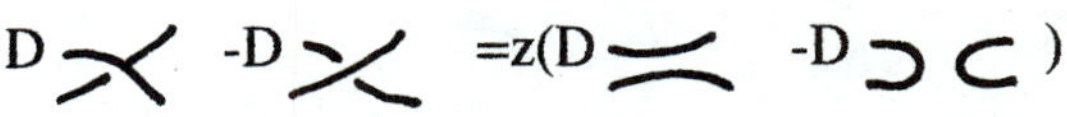

D is normalized to an ambient isotopy invariant G via the equation

$$G_K = a^{-w(K)} D_K$$

The loop variable for D and G is

$$\mu = 1 + (a-a^{-1})/z$$

so that

$$D_{O\,K} = \mu D_K.$$

For the record, the translation between D and L is given by the equations:

$$D_K(a,z) = (-1)^{c(K)-1}(i^{w(K)})L_K(-ia,iz) \quad , \quad (i=\sqrt{-1})$$

where i denotes the square root of minus one, and c(K) is the number of link components of K.

One purpose of the knot-theoretic part of this paper is to show how naturally certain solutions to the Yang-Baxter equations appear in the course of constructing models for these polynomials.

4. The Bracket Polynomial

This section is a brief introduction to the bracket polynomial ([33],[34],[37],[39],[40],[42],[43],[44]). The principles involved in construction of the bracket will be used in forming other state models. The *bracket polynomial*, <K>(A,d>, is a polynomial in commuting variables A and d associated to a link diagram K. At the outset this polynomial is well-defined on diagrams taken up to planar isotopy. It is not invariant under the Reidemeister moves without an appropriate specialization of the variables. In this section we shall see that with one specialization, the bracket yields a construction of the original Jones polynomial and with another specialization it gives a version of the Potts model in statistical mechanics.

The bracket is defined by the recursion formula

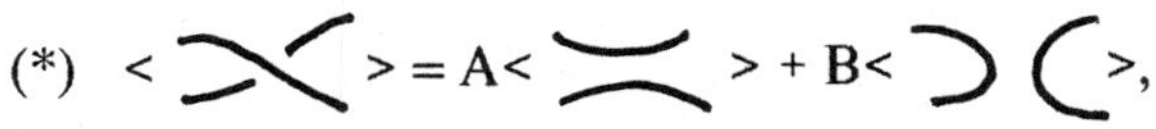

and the rules

$$\langle K \bigcirc \rangle = d\langle K \rangle,$$

$$\langle \bigcirc \rangle = 1.$$

where K 0 denotes a diagram K with an extra Jordan curve component,0, placed anywhere in the regions of the diagram.

This association of the labels A and B to the two modes of splitting the unoriented crossing is well-defined because an unoriented crossing distinguishes two out of the four local regions via the convention shown below.

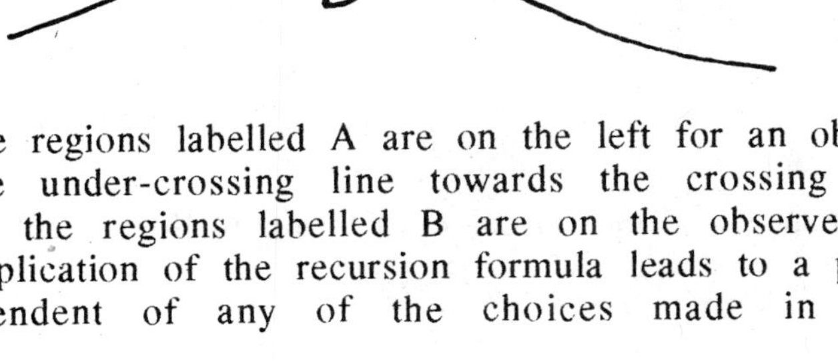

The regions labelled A are on the left for an observer walking along the under-crossing line towards the crossing (either way). Similarly, the regions labelled B are on the observer's right.

Application of the recursion formula leads to a polynomial that is independent of any of the choices made in the recursive process.

State Summation for the Bracket

An explicit formula for <K> is obtained as follows. Call a *state* *S* of K any one of the (two raised to the number of crossings) smoothings of the diagram into a collection of simple closed curves in the plane. Such a smoothing entails a two-fold choice at each crossing site.

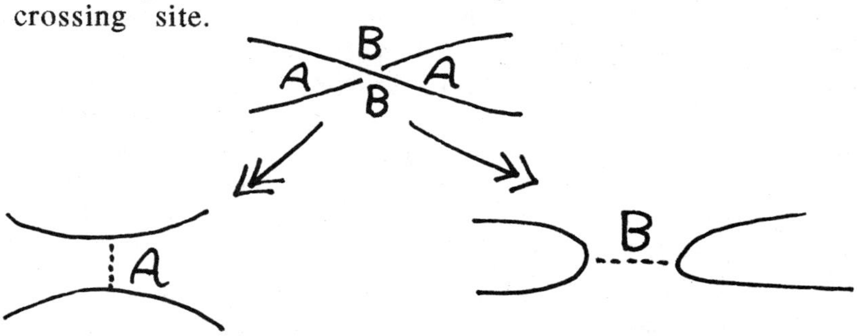

For each smoothed site, define its *vertex weight* by the formulas

Let <K | S> denote the product of all the vertex weights for a given state S relative to the diagram K.

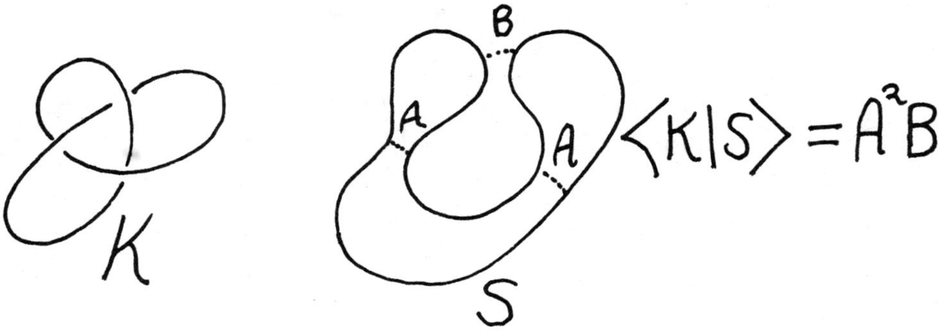

For each state S, define the *norm* of the state, S , by the formula

$$||S|| = |S| - 1$$

where |S| denotes the number of simple closed curves in the state S.

Then

$$<K> = \sum_S <K|S>\, d^{||S||}$$

gives an explicit formula for the bracket as a sum of products of vertex weights (augmented by a global evaluation of the state). One can take this equation as the definition of the bracket. The recursion formulas follow directly from it.

This state summation has a close relationship with models in statistical mechanics. See [40] for a detailed comparison of the bracket with the Potts model.

The Topological Bracket

With the rule

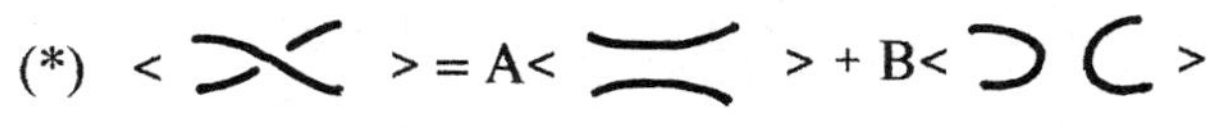

it is easy to see that

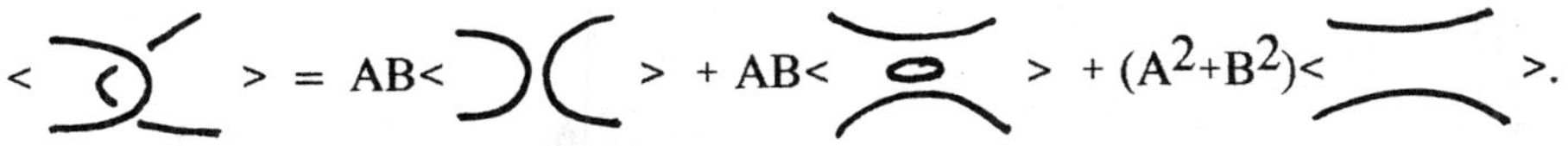

Hence, in order for <K> to be invariant under the type II move, we need that

AB=1 and

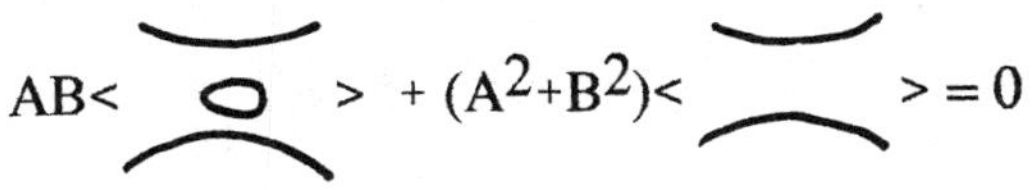

or, equivalently, that

$$B=A^{-1} \quad \text{and}$$

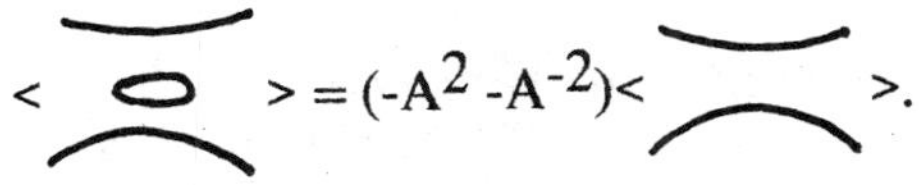

Letting d denote this factor

$$d = -A^2 - A^{-2},$$

we then have that if S is a collection of n disjoint closed curves in the plane, then

$$<S> = d^{n-1}.$$

Invariance under move III follows automatically from the equation (*) and type II invariance:

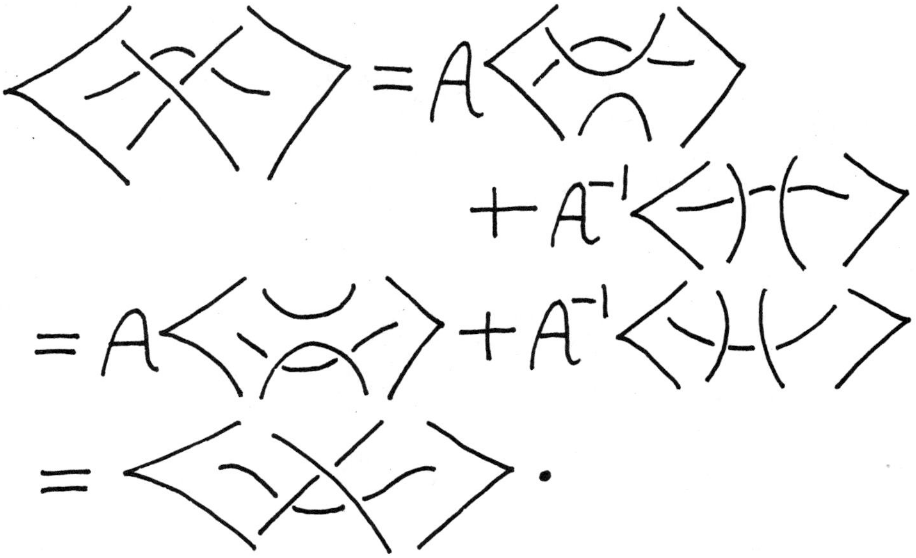

By writhe-normalizing, we define a polynomial

$$f_K = (-A^3)^{-w(K)}\langle K\rangle$$

defined for oriented links K. f_K is an invariant of ambient isotopy. This gives a model for the original Jones polynomial. The precise result is:

Theorem [34]. *Let* $V_K(t)$ *denote the original Jones polynomial. Then*

$$V_K(t) = f_K(t^{-1/4}).$$

Note that $\langle K\rangle$ is a special case of the polynomials L and D discussed in the previous section. From this it follows that the original Jones polynomial is a special case of the Kauffman polynomial.

We conclude this section with two examples that are very closely related to the topological bracket.

Example 1. The *dichromatic polynomial*, $Z_G = Z_G(v,q)$, of a plane graph G is defined by the the following formulas.

$$Z_G = Z_{G'} + vZ_{G''}$$
$$Z_{*G} = qZ_G$$

where G' is the result of deleting an edge e of G, G" is the result of contracting this same edge to a point and *G denotes the disjoint union of G with a single vertex (*). The dichromatic polynomial is a polynomial in the commuting variables v and q. When v=-1 and q a positive integer, the polynomial Z_G gives the number of proper colorings of the graph G with q colors. (A coloring is said to be proper if adjacent vertices receive different colors. Two vertices are adjacent if they share an edge in G.)

In [37] and [40] I show that the dichromatic polynomial of a plane graph can be computed by using a bracket expansion on link diagrams (a non-topologically invariant bracket). This result has the form

$$Z_G = q^{N/2}\{K(G)\}$$

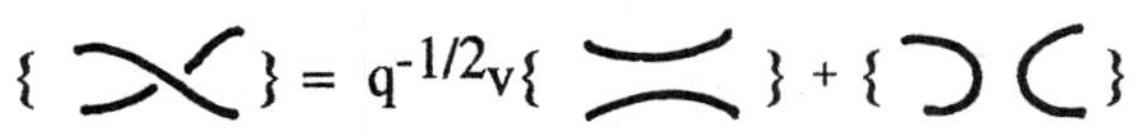

$$\{O\ K\} = q^{1/2}\{K\}$$

$$\{O\} = q^{1/2}$$

With these settings, Z_G computes the dichromatic polynomial for a plane graph G when K=K(G) is taken to be the alternating link uniquely associated to G via the formation of the medial graph for G.

Here is a quick description of the medial graph and the alternating link K(G). Given a link diagram K we can associate to it a plane graph G(K) by the following algorithm: The link diagram has an associated *shadow* that is a 4-valent plane graph Sh(K). The shadow is obtained by replacing each crossing by a 4-valent vertex as shown below.

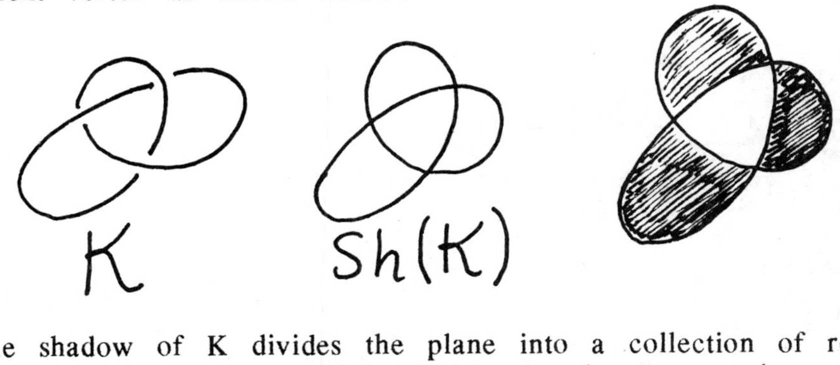

The shadow of K divides the plane into a collection of regions. Color these regions with two colors so that two regions sharing an edge of the shadow are colored differently. This can always be done (exercise!). Call the two colors *black* and *white*. Refer to the black regions as the *shaded* regions. Let the unbounded region in the plane be colored white. This makes the corresponding coloring unique.

Let G(K)=G(SH(K)) be the graph with one vertex for each shaded region in this coloring and one edge for each crossing shared by shaded regions. This includes loops for crossings shared by a region with itself.

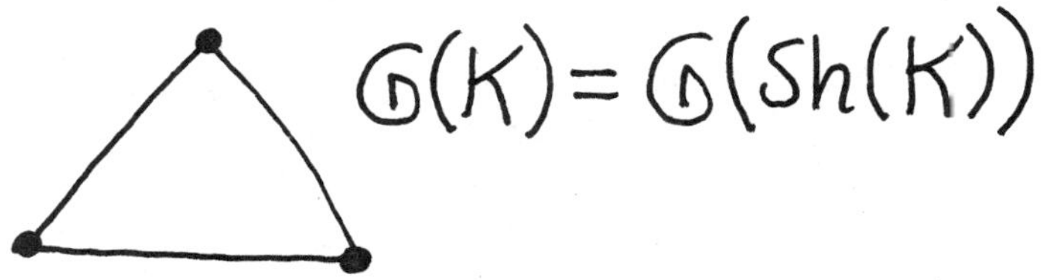

Given an arbitrary connected plane graph G, the medial, M(G), is a 4-valent plane graph such that G(M(G))=G. This inverse process is indicated below. M(G) is obtained by curves parallel to the boundaries of the regions of the plane graph G, and then putting in 4-valent vertices between each pair of parallel arcs along an edge of G as shown below.

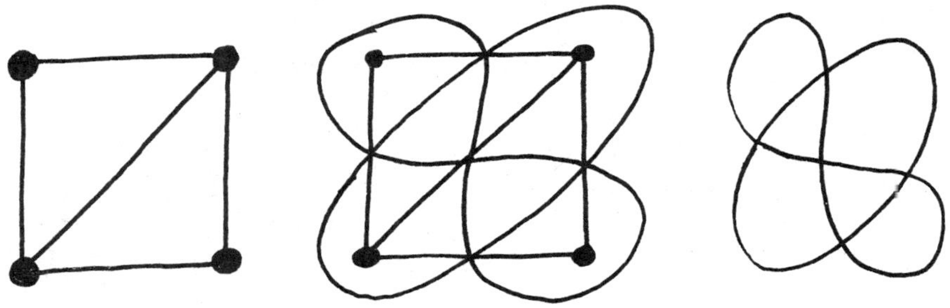

Finally, the knot diagram K(G) is obtained from M(G) by making each 4-valent vertex into a crossing so that each crossing is of "A-type" with respect to the shading. See the illustration below. It is an easy exercise to verify that the resulting link, K(G), is alternating in the sense that the crossings alternate between over and under as one traverses the diagram.

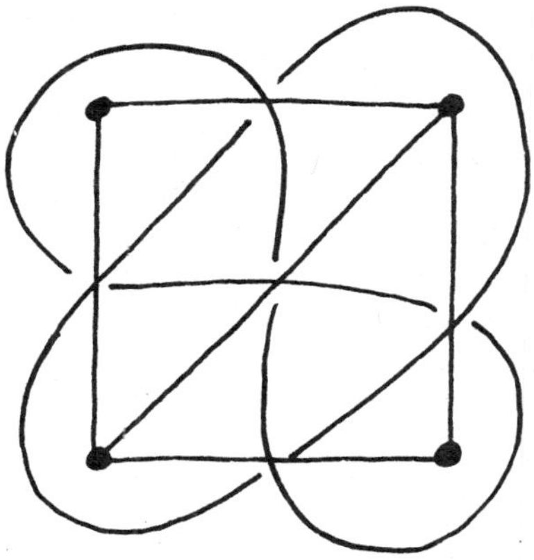

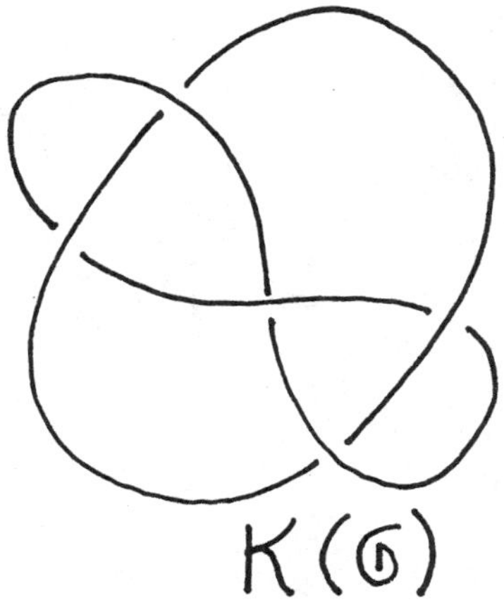

Example 2. The Potts Model.

Example 1 is directly related to the Potts model in statistical mechanics. In the Potts model each vertex of a plane graph G can receive one of q (a positive integer) possible "colors".

Here color is a neutral term for a local physical state. For example, there could be little magnets at each site (vertex) of the graph G, and two possible orientations of each magnet (q=2). Adjacent sites might have their magnets aligned or oppositely alligned. In one physical interpretation the magnets do not like to be alligned. They like to have opposite poles adjacent. In this case the lowest energy states of the system are those with the least number of coincidences of adjacent colors.

A state σ of G is defined to be an assignment of colors to the vertices of G. The *energy*, $E(\sigma)$, is defined to be the number of coincidences of colors. $E(\sigma)$ is equal to the number of edges of G such that the endpoints of the edge receive the same color. From the point of view of graph-coloring, $E(\sigma)$ is a measure of the deviance of s from being a proper coloring of G. The proper colorings of G are the states σ such that $E(\sigma)$ is zero. It is natural in statistical mechanics to associate with a physical system its partition function, Z_G, defined to be the summation over all the states σ of the system the weights $p(\sigma) = \exp((-1/kT)E(\sigma))$:

$$Z_G = \sum_\sigma \exp((-1/kT)E(\sigma)).$$

Each weight $p(\sigma)$ is proportional to the probability that the system is in the state σ. Here T denotes the temperature of the system and k is a constant (Boltzmann's constant). Thus the partition function yields the actual probability of states, $P(\sigma)$, via the equation $P(\sigma) = p(\sigma)/Z_G$. Many other physical quantities related to the thermodynamics of the system are obtained from the partition function. If at certain temperatures the physical system undergoes phase transition (e.g. from a liquid to a gas), then one expects that there will be a discontinuous change in the partition function.

The Potts model is an idealization of a real physical system. Nevertheless, in the case q=2 (the Ising model) it was proved (See [7].) in the 1940's by Onsager and then others that there is a phase transition in this model for G a limit of rectangular planar lattices. Similar results have been obtained for other cases of q in the Potts model, but the story is not over yet. For these reasons, the Potts model is of great interest to statistical mechanics.

In [64] Temperley and Lieb proved the remarkable result that the Potts partition function is a dichromatic polynomial. They showed that

$$Z_G = Z_G(v,q) \quad \text{where } v = \exp(-1/kT) -1.$$

Once one sees this fact, it is easy to prove and so we leave it as an exercise for the reader! Armed with the fact that the Potts partition function is a dichromatic polynomial, we now see that for a plane graph, the Potts partition function can be computed as a bracket polynomial from the associated alternating medial K(G).

The bracket expansion is not just analogous to a partition function in statistical mechanics is actually *is* a partition function for appropriate choices of variables.

5. Abstract Tensors and the Yang-Baxter Equation

When $M^i{}_j$ and $N^i{}_j$ denote tensors or matrices it is customary for a repetition of an upper and lower index to denote summation over the range of values for that index. This is called the *Einstein summation convention.*

Ordinary matrix multiplication is the simplest example of the convention:

$$(MN)^i{}_j = M^i{}_k N^k{}_j = \sum_k M^i{}_k N^k{}_j$$

The trace of a matrix with one upper index and one lower index is given by writing the same index label on the top and on the bottom:

$$\mathrm{tr}(M) = M^i{}_i = \sum_i M^i{}_i \,.$$

A diagrammatic form of the Einstein summation convention has been used by various authors (e.g. [57]).
In this system we

1. Replace an object with indices by a diagram (a box, a triangle, ...) with labelled lines corresponding to these indices.

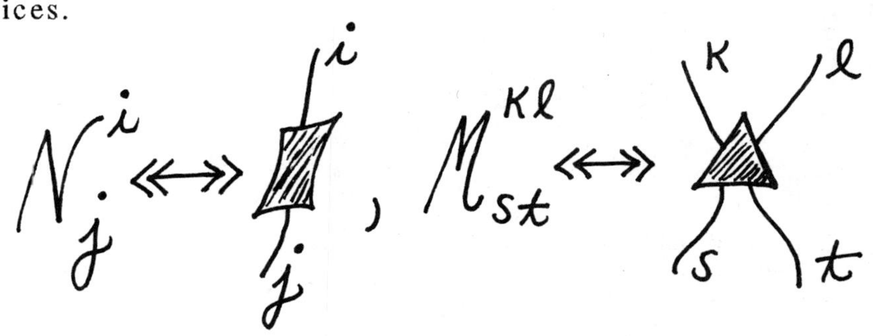

2. Tie those lines together whose indices are repeated.

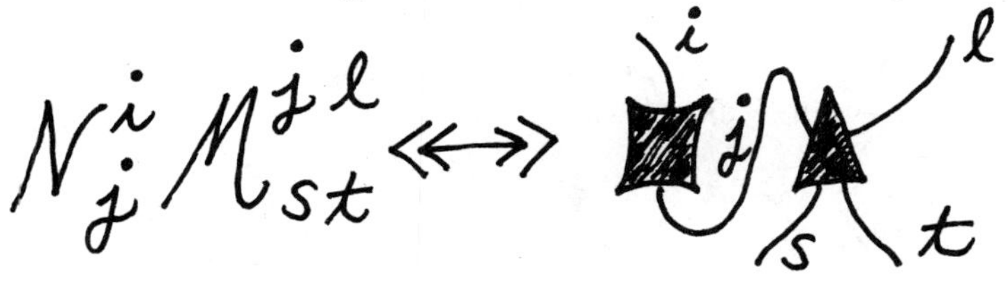

If all upper indices are tied to lower indices, then the form has contracted to a scalar. Thus, if

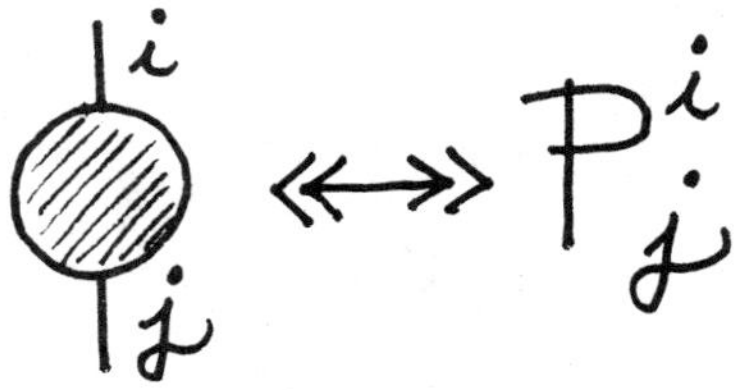

is a matrix, then

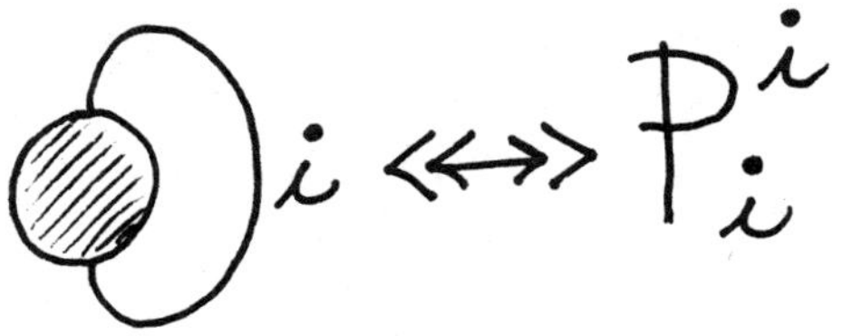

is its trace.

Remark. Later we shall have occasion to use curved parts of the connecting lines as matrices. Thus we shall have *cups* and *caps:*

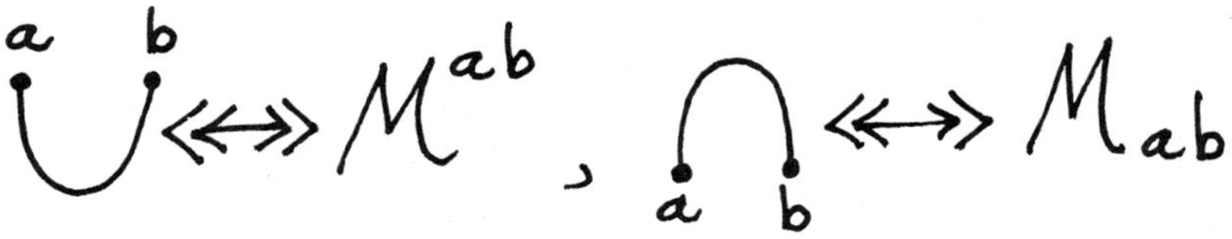

These cups and caps will then modify the trace by their presence in the diagrammatic. For the present we follow the convention that *any line not specifically marked as a matrix is a Kronecker delta* (See discussion below).

This mode for dealing with tensor formalism makes possible the direct illustration of the transition between a link diagram and the corresponding state model.

We shall associate matrices to the crossings of a link diagram. Thus, with an assignment as shown below

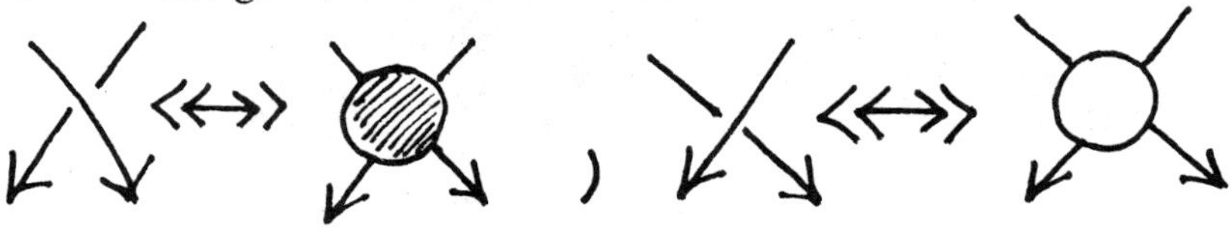

any oriented link diagram corresponds to a contracted scalar via the summation convention.
For example

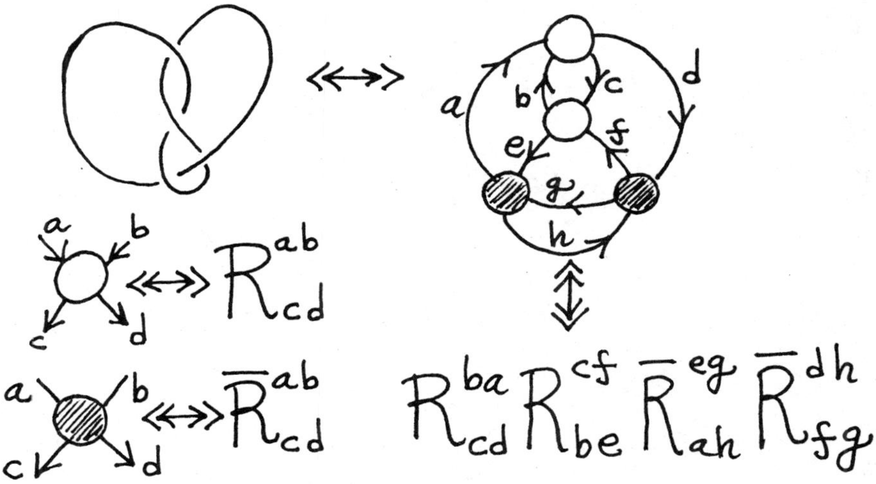

$$\longleftrightarrow R^{ab}_{cd} \qquad \longleftrightarrow \overline{R}^{ab}_{cd}$$

$$R^{ba}_{cd}\, R^{cf}_{be}\, \overline{R}^{eg}_{ah}\, \overline{R}^{dh}_{fg}$$

One can then ask for general conditions on these matrices to insure the topological invariance of the associated scalar. For example, the type II move

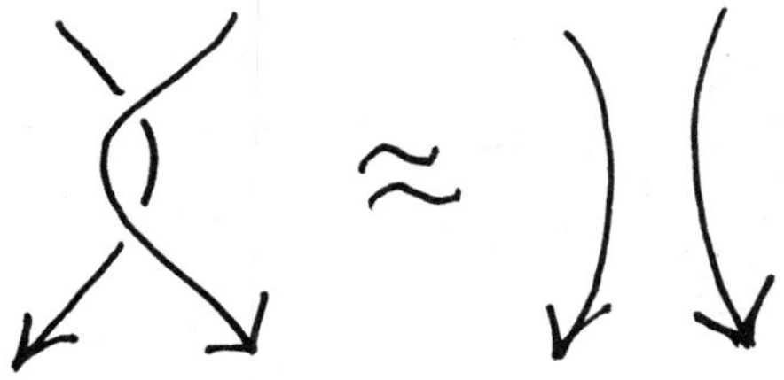

translates to the equivalence

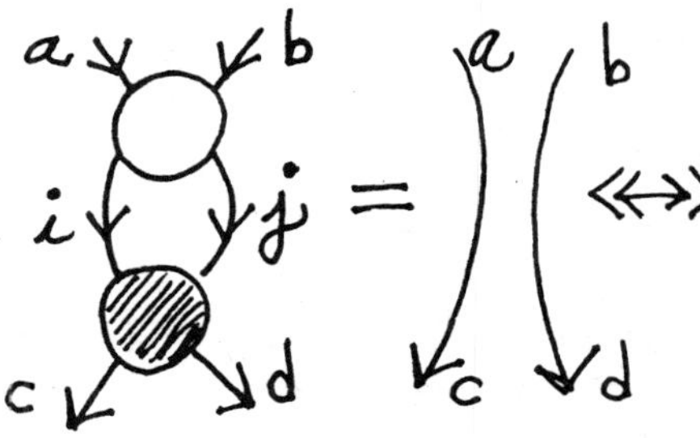

$$\longleftrightarrow R^{ab}_{ij}\, \overline{R}^{ij}_{cd} = \delta^a_c\, \delta^b_d .$$

We identify a line segment with indices at its endpoints as a Kronecker delta:

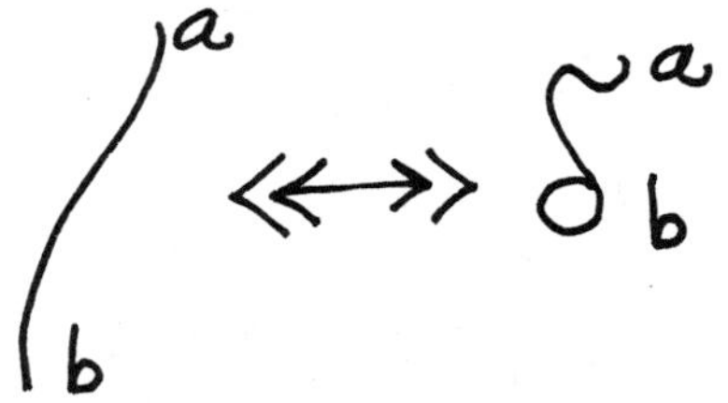

Thus the type II move gives the condition

$$R^{ab}{}_{ij}(R^{-1})^{ij}{}_{cd} = \delta^a{}_c \delta^b{}_d.$$

The matrices associated with positive and negative crossings must be inverses of one another. Call this condition *unitarity.*

In a similar way, the reverse oriented type II move translates to an equation that I shall call *cross channel unitarity:*

$$(R^{-1})^{ia}{}_{jb} R^{jd}{}_{ic} = \delta^a{}_c \delta^b{}_d.$$

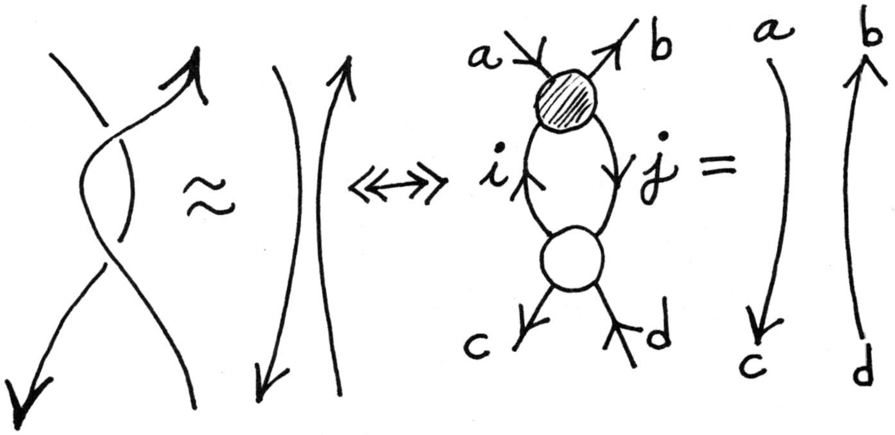

In practice, we shall not always assume that the matrices satisfy cross channel unitarity. Instead, extra structure in the state model will insure the invariance under the reversed type II move. Nevertheless the pattern of the cross-channel connection must be considered.

Finally, the third move translates as shown in Figure 9.

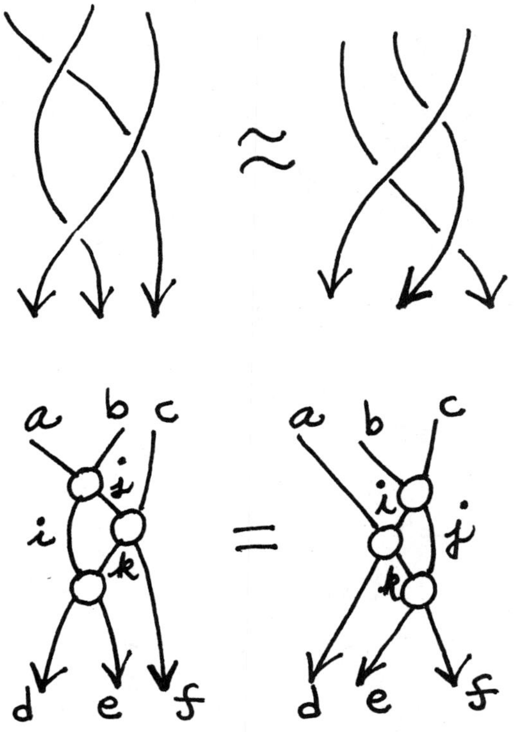

$$\mathbf{R^{ab}{}_{ij}R^{jc}{}_{kf}R^{ik}{}_{de} = R^{bc}{}_{ij}R^{ai}{}_{dk}R^{kj}{}_{ef}}$$

Quantum Yang-Baxter Equation
(without rapidity parameter)
Figure 9

The algebraic condition

$$R^{ab}{}_{ij}R^{jc}{}_{kf}R^{ik}{}_{de} = R^{bc}{}_{ij}R^{ai}{}_{dk}R^{kj}{}_{ef}$$

is (ignoring a certain physical parameter) the *Quantum Yang-Baxter Equation* (QYBE). (See [72].).

Definition 5.1. *For a link diagram K, let <K;R> denote the scalar that is obtained by replacing each crossing of L by R or its inverse , and taking the generalized trace as described above (summing over all assignments of indices to the lines of the diagram).*

Theorem 5.2. *Let R be a matrix satisfying*
1. unitarity - $RR^{-1} = 1$
2. cross-channel unitarity - (see above)
3. quantum Yang-Baxter equation
Then <K;R> is an invariant of regular isotopy of the link K.

Proof. The theorem follows at once from the discussion that precedes it.

Example. Let R be given by the formula below with indices in the set $\{1,2,...,q\}$ and A and B $(B=(1/A))$ given by the formulas:

$$R^{ab}_{cd} = A\delta^{ab}\delta_{cd} + A^{-1}\delta^{a}_{c}\delta^{b}_{d}$$

$$(q=-A^{-2}-A^{2}).$$

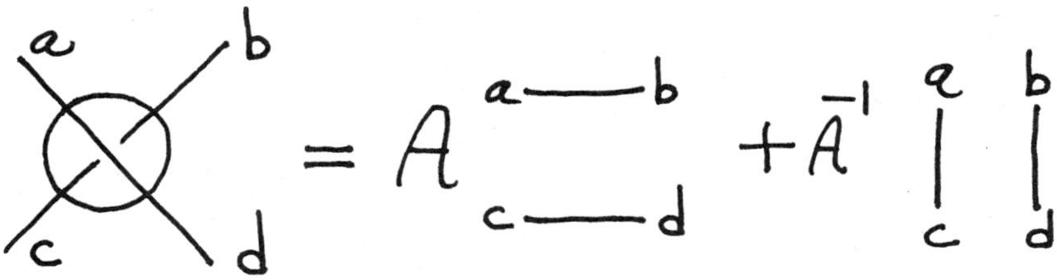

It is easy to see that R satisfies all three conditions of the theorem. We have already done the work for this model in constructing the bracket polynomial in section 4. <K;R> gives a specialization for the bracket where the loop value is the integer q. Note that we have set the element A so that $q = -A^{-2} - A^{2}$. In this case <K;R> is defined for unoriented diagrams (corresponding to the usual conventions for the bracket.)

Topological Remark. By the results of Section 2, we know that it is sufficient for regular isotopy invariance (in the presence of both type two moves) to have invariance under the moves of type IIIA in positive and negative orientation. This is exactly the invariance given by the QYBE for the matrix R (associated to positive crossings) and the inverse of R associated to negative crossings. In the presence of invariance under type II oriented and reversed, one then gets (see section 2) invariants of regular isotopy. Thus in constructing state models we shall use the QYBE for type II invariance and the main technical problem will be to obtain invariance under the reversed type two move.

If one simply wants a representation of the n-strand braid group, then a solution R and R^{-1} to QYBE is sufficient. See [5] and [18] for information about using QYBE to create invariants of links via Markov trace and braid-group representation.

Rapidity Remark. The R-matrix often includes a "rapidity parameter" (corresponding to momentum difference in the scattering context). In section 9 we will show how to construct state models that take account of such a parameter in a limited sense. In general this approach to QYBE ignores the dependence on such a parameter, or makes the dependence as simple as possible. There may be subtler ways to include the rapidity structure in the topology -- this is an open question.

Physical Remark. The Quantum Yang-Baxter Equation has been studied extensively in a number of contexts in mathematical physics. It appears in classical (non-quantum) statistical mechanics in Baxter's work on exactly solvable models - as a condition for transfer matrices to commute with one another.

The quantum mechanical interpretation corresponds to thinking of the matrix R as an amplitude (complex probability amplitude) for a scattering process for two particles with incoming "spins" a and b , and outgoing spins c and d. (The spins correspond to the indices of the matrix. These indices will be regarded as chosen from a subset of the real numbers for the rest of this discussion.) In this picture one can think of the graphical crossing in the link diagram as indicating a single scattering event with a local arrow of time that is coincident with the orientation of the diagram.

The scattering picture corresponds to quantum mechanics because the rule for the production of amplitudes from a process made out of two-particle events corresponds exactly to our diagrammatic matrix multiplications. The amplitude for a complex process with a given set of input spins and output spins is obtained by summing over the products of amplitudes for all choices of intermediate spins.

6. Bracket Model and Quantum Groups

An Unoriented Model

Here is a direct method to capture the full bracket polynomial as an abstract tensor model. Arrange the link diagram so that its only critical levels with respect to a choice of vertical are maxima, minima and crossings. The diagram then decomposes into maxima, minima and crossings and we shall associate a matrix with each of these as indicated below.

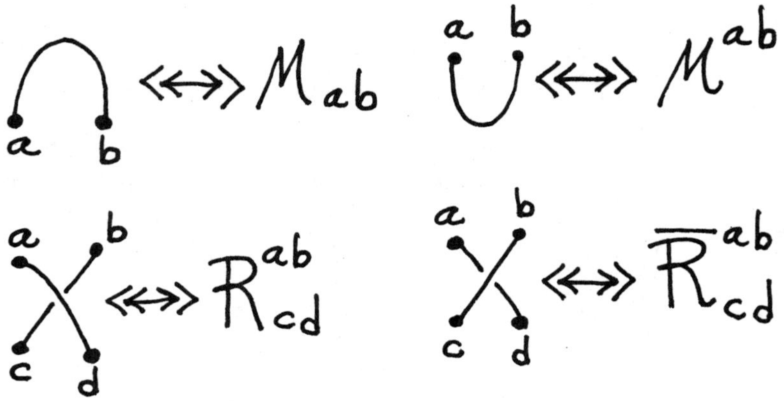

Specifically we shall take M as shown below with $M^{ab} = M_{ab}$.

$$M = \begin{bmatrix} M_{11} & M_{12} \\ M_{21} & M_{22} \end{bmatrix} = \begin{bmatrix} 0 & \sqrt{-1}A \\ -\sqrt{-1}A^{-1} & 0 \end{bmatrix}.$$

This means that $M^2=I$. Note that this choice corresponds to the topological move shown below - cancellation of adjacent maxima and minima.

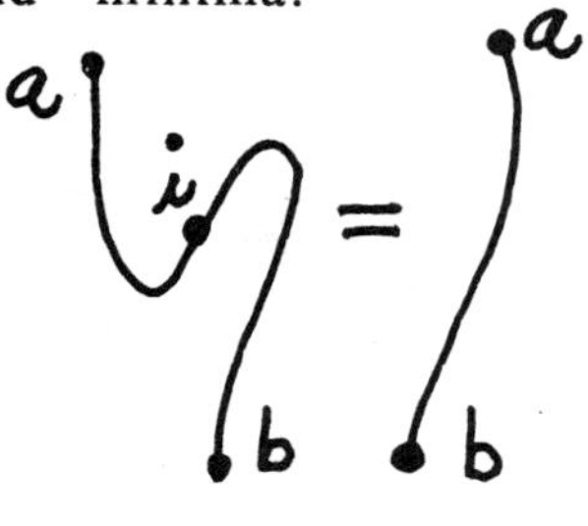

$$: M^{ai} M_{ib} = \delta^a_b.$$

Note also that the value of the loop in this model is equal to $(-A^2-A^{-2})$.

$$= M_{ij}M^{ij} = \sum_{ij}(M_{ij})^2$$

$$= -A^2 - A^{-2}.$$

Now we take

$$R^{ab}{}_{cd} = AM^{ab}M_{cd} + A^{-1}\,\delta^a{}_c\delta^b{}_d.$$

This corresponds directly to the diagrammatic

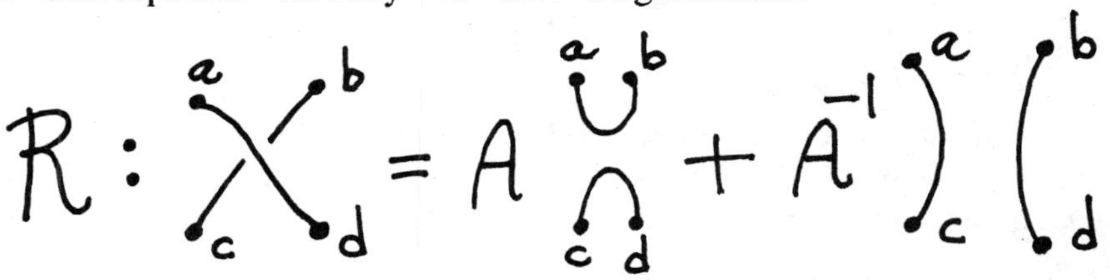

and it is not hard to verify directly that $R^{ab}{}_{cd}$ satisfies the Yang-Baxter equation. You can do this by manipulating diagrams! The resulting model produces the bracket state sum since it satisfies the axioms for the bracket. This solution to the Yang-Baxter equation is related to the quantum group associated with SL(2).

Now look at R more explicitly: $\mathbf{R=AU + A^{-1}I}$.

$$U = \left(M^{ab}M_{cd}\right) = \begin{bmatrix} 0 & 0 & 0 & 0 \\ 0 & -A^2 & 1 & 0 \\ 0 & 1 & -A^{-2} & 0 \\ 0 & 0 & 0 & 0 \end{bmatrix}$$

$$\Rightarrow A^{-1}R = \begin{bmatrix} A^{-2} & 0 & 0 & 0 \\ 0 & -A^2+A^{-2} & 1 & 0 \\ 0 & 1 & 0 & 0 \\ 0 & 0 & 0 & A^{-2} \end{bmatrix}.$$

We see that R can be described by the formalism

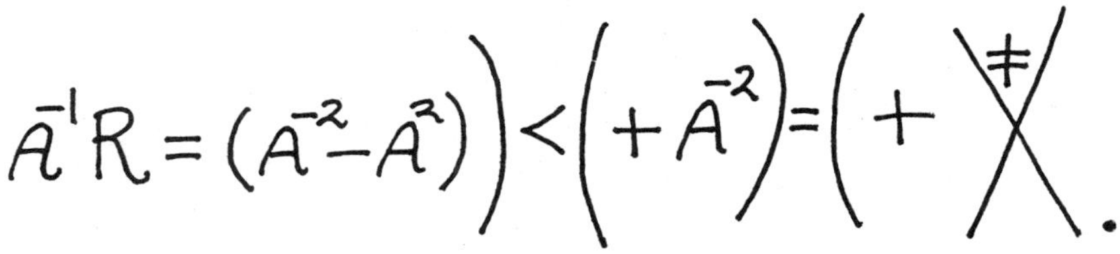

where each line can be assigned indices ranging from 1 to 2 and an inequality between lines means that they are labelled differently.

In this form, there is an obvious generalization of R to a matrix of the form

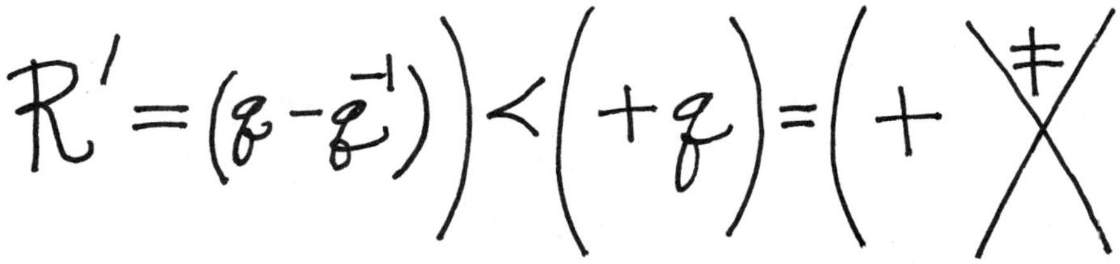

for an arbitrary ordered index set. As we shall see in the next section, this new R is still a solution to the Yang-Baxter equation and it leads to new invariants - specializations of the Homfly polynomial.

The model for the bracket that emerges is as follows: Choose a Morse function on the link diagram K so that it is decomposed into maxima, minima and crossings. With respect to this decomposition write the diagram as a product of abstract tensors

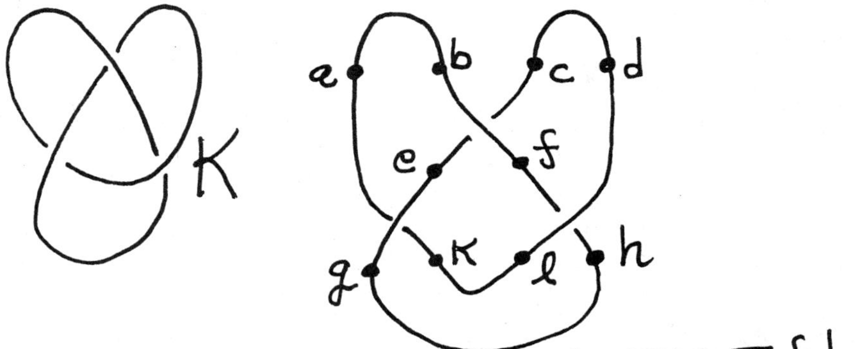

$$T(K) = M_{ab}M_{cd}M^{kl}M^{gh}R^{bc}_{ef}\bar{R}^{ae}_{gk}\bar{R}^{fd}_{lh}.$$

Call this product T(K).

If we interpret the abstract tensors to be explicitly given as above, then $<K> = T(K)/d$ where d is the value of the loop.

In general , it is easy to see that if M^{ab} , M_{cd} , R^{ab}_{cd} and $\overline{R}^{ab}_{cd}$ are any matrices so that M^{ab} and M_{cd} are inverses, R^{ab}_{cd} and $\overline{R}^{ab}_{cd}$ are inverses and each satisfy the QYBE, *and* so that the following *twist condition* holds

$$(R^{-1})^{ab}_{cd} = M_{ci}R^{ia}_{dj}M^{jb} = M^{ai}R^{bj}_{ic}M_{jd},$$

then T(K) is an invariant of regular isotopy of K.

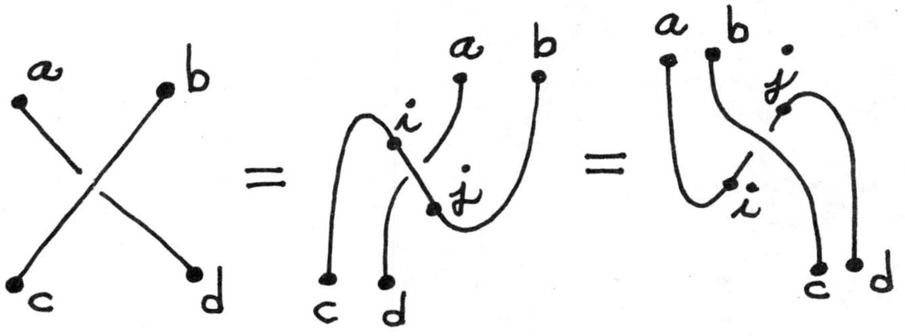

Furthermore , if we take symbolic tensors with just these properties, then we obtain an abstract tensor algebra that completely classifies unoriented links up to regular isotopy. This is general nonsense, but a good way to think about the structure of the model.

Braids

It is worth noting how this symbolic tensor model works in the case of braids. The braid itself has no maxima or minima. The closure of the braid is obtained by connecting braid inputs to braid outputs by arcs , each arc having one maximum and one minimum in the form

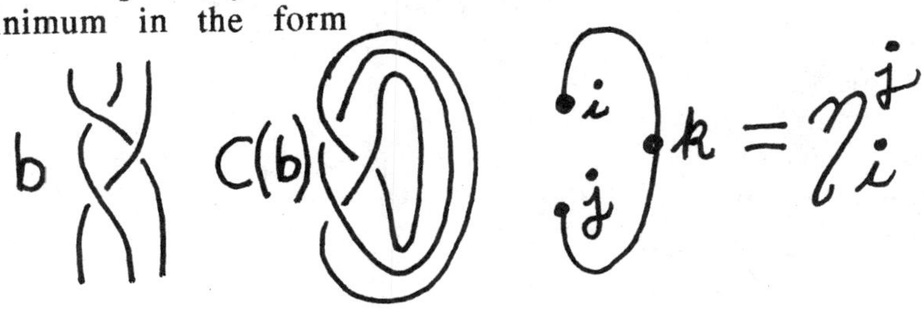

This results in the product

$$\eta^j{}_i = M_{ik}M^{jk}, \qquad \eta = MM^T.$$

Thus, it follows at once that the bracket on the closure C(b) of a braid b is given by the formula shown below

$$< C(b) > = \mathrm{trace}(\tau\ \rho(b))$$

where $\tau = \eta \otimes \eta \otimes \eta \ldots \eta \otimes \eta$, with n factors in the tensor product and $\rho(b)$ denotes the representation of the braid obtained from the R-matrices. (Compare [40].)
In the case at hand we have the upper and lower M-matrices identical to each other and the matrix η will correspond to the product of M with its transpose.

Quantum Groups

The "generator" of the bracket is the matrix **E**,

$$E = \begin{bmatrix} 0 & A \\ -A^{-1} & 0 \end{bmatrix}$$

a deformation of the epsilon ε

$$\varepsilon = \begin{bmatrix} 0 & 1 \\ -1 & 0 \end{bmatrix}$$

I say that this deformed epsilon generates the bracket because the R-matrix for the bracket is constructed directly from it. This is related to the fact that the Jones polynomial is obtained from a representation of the braid group to the Temperley Lieb algebra rather than a Hecke algebra. See [40] for a discussion of this point using the deformed epsilon.

The epsilon is an invariant of the group SL(2): If P is a matrix with commutative and associative entries, then

$$P\varepsilon P^T = \mathrm{DET}(P)\varepsilon.$$

$$SL(2) = \{ P \mid \mathrm{DET}(P)=1 \}=\{P \mid P\varepsilon P^T = \varepsilon\}.$$

This leads directly to the question:

Can SL(2) be generalized to leave the deformed epsilon invariant?

Consider a matrix P that leaves the deformed epsilon invariant. (We assume that the elements of P belong to a ring whose elements commute with A - the non-trivial entry of E). That is,

$$P E P^T = E$$
$$\text{and}$$
$$P^T E P = E.$$

It is easy to see that this invariance is equivalent to the conditions

$$ba = A^2 ab \qquad ca = A^2 ac$$
$$dc = A^2 cd \qquad db = A^2 bd$$

$$bc = cb$$

$$ad - da = (A^{-2} - A^2)bc$$

$$ad - A^{-2}bc = 1$$

when P is the matrix shown below.

$$P = \begin{pmatrix} a & b \\ c & d \end{pmatrix}$$

These are exactly the defining relations for the quantum group $SL(2)_q$ with $A=\sqrt{q}$ in Manin's formulation (See [14],[55].).

A further ingredient in a quantum group **Q** is a mapping $s:Q \text{ -----}> Q$ that is an antimorphism ($s(ab)=s(b)s(a)$). The antipode is an analog of the existence of inverses in a group. In the case of $SL(2)_q$ the antipode applied to the matrix P gives the matrix s(P) shown below and it is an interesting exercise to verify that P and s(P) are (left and right) inverses of each other.

$$s(P) = \begin{pmatrix} d & -qb \\ -q^{-1}c & a \end{pmatrix}, \quad q = A^2.$$

Dualizing

$SL(2)_q$ is often described dually as a deformation of the Lie algebra sl(2). See Drinfeld [14] for the specific translation. The dual algebra is called the *quantum universal enveloping algebra (QUEA) of sl(2)* $U_h(sl(2))$. It is given via generators and relations

$$[H\ ,\ X] = 2X^+$$
$$[H\ ,\ X^-] = 2X^-$$
$$[X^+,X^-] = \sinh((h/2)H)/\sinh(h/2)$$
$$(\ [A,B] = AB-BA\ ,\quad q = e^h)$$

as a deformation of sl(2) where in sl(2) the elements H, X^+ and X^- are given explicitly as shown below.

$$H = \begin{pmatrix} 1 & 0 \\ 0 & -1 \end{pmatrix},\ X^+ = \begin{pmatrix} 0 & 1 \\ 0 & 0 \end{pmatrix},\ X^- = \begin{pmatrix} 0 & 0 \\ 1 & 0 \end{pmatrix}$$

The quantum universal enveloping algebra is a Hopf algebra via the co-multiplication

$$\Delta(H) = H \otimes 1 + 1 \otimes H$$

$$\Delta(X^{\pm}) = X^{\pm} \otimes e^{hH/4} + e^{-hH/4} \otimes X^{\pm}$$

One can form the dual by taking the coordinate functions in the representation ρ: $U_h sl(2)$ -----> $M_2(C[[h]])$

$$\rho(H) = \begin{pmatrix} 1 & 0 \\ 0 & -1 \end{pmatrix},\ \rho(X^+) = \begin{pmatrix} 0 & 1 \\ 0 & 0 \end{pmatrix}$$

$$\rho(X^-) = \begin{pmatrix} 0 & 0 \\ 1 & 0 \end{pmatrix},\ \rho = \begin{pmatrix} \rho_{11} & \rho_{12} \\ \rho_{21} & \rho_{22} \end{pmatrix}$$

$$\rho_{ij} : U_h sl(2) \text{ -----> } C[[h]].$$

Given two functions f and g on $U_h sl(2)$, we define their product fg via the co-algebra structure:

$$fg(x) = f \otimes g(\Delta(x)).$$

Letting the entries of ρ correspond to the elements **a,b,c,d,** with this multiplication, we obtain $SL(2)_q$ as the dual of $U_h sl(2)$.

Conclusion

This completes a sketch of the connection between diagrammatic and knot theoretic ideas and the construction of the quantum group. We have seen the quantum group arise naturally from the question of the symmetry of a tensor whose deformation is motivated by topological considerations. Quantum groups contain solutions to the Yang-Baxter equation and can be used to construct link invariants. See [61],[59],[44.8],[44.9].

7. Yang-Baxter Models for Skein Polynomials - Oriented Models

I call the models discussed in this section *4-local* because they depend upon four local state configurations at an oriented crossing. The four local configurations will be denoted symbolically by the notations shown below:

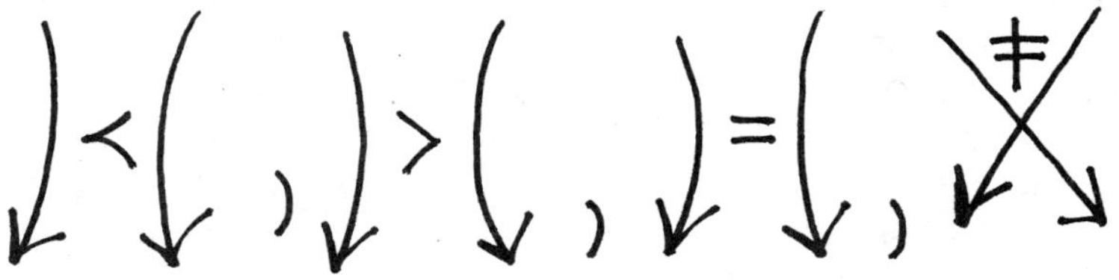

There are four potential distinctions available at an oriented crossing. These distinctions are related to either splicing the crossing, or to projecting it (so as to ignore its over/under crossing structure). In a spliced crossing, the two segments are labelled with spins a and b -- giving three possible relations: $a<b, a>b, a=b$. This , together with the switching of distinct a and b (hence the sign =) in the crossed state provides the 4-fold distinction.

Following the ideas of the previous section, we shall construct a state model that is based on one of the simplest solutions to the Quantum Yang-Baxter Equation. Since the first model desired is a model for the regular isotopy version of the Homfly polynomial (see section 3) , we want the state summation [K] to satisfy the exchange identity

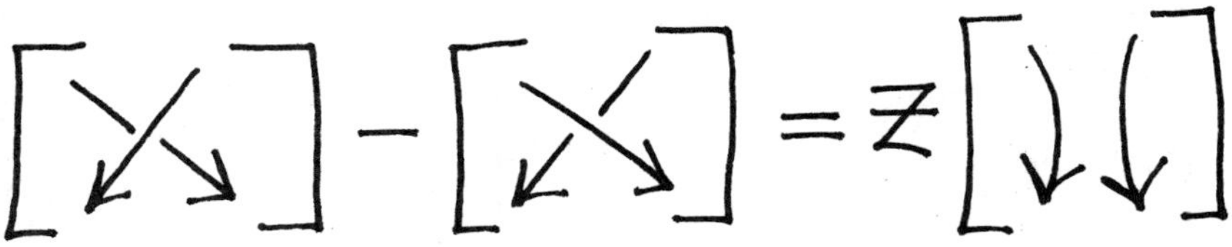

The analog of this condition for the Yang-Baxter matrix is the equation

$$\mathbf{R} - \mathbf{R}^{-1} = \mathbf{z}\mathbf{I}$$

Now suppose that R is chosen so that it is *spin preserving* , that is,

$$\mathbf{R}^{ab}{}_{cd}$$

is non-zero only if the sum of the input spins is equal to the sum of the output spins : a+b = c+d. (In section 10 we shall see the clearest theoretical reason for putting this restriction on the R-matrix.)

The simplest way to accomplish this requirement is when *the set of output spins is equal to the set of input spins.* In this case , the local possibilities are as shown in Figure 11. There are three cases of "parallel interaction" for the spins a<b , a>b , and a=b. There is one possibility for a "crossover" with a and b unequal - to avoid duplication.

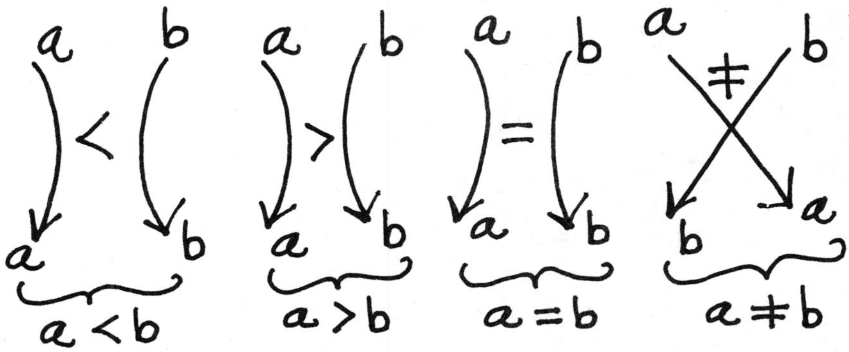

Simple Spin-Preserving Configurations
Figure 11

The symbolism of Figure 11 can be used in one or all of the following ways:

1. to indicate the relative positions of the scattered particles,
2. to indicate a local splitting or projection (in the case of the crossover) of a link diagram,
3. as a diagrammatic tensor that is equal to one if the indicated condition is met, and is zero otherwise. Thus the tensor formalism is given by the conditions below:

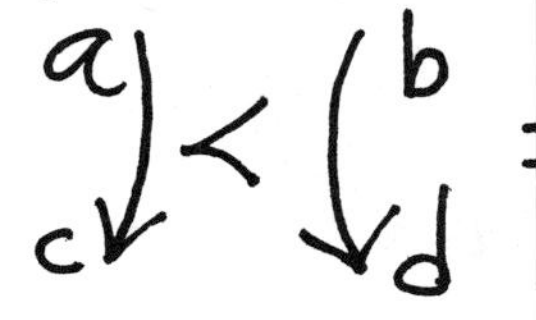

$$= \begin{cases} 1 & \text{if } a=c, b=d, \\ & \quad a<b. \\ 0 & \text{otherwise.} \end{cases}$$

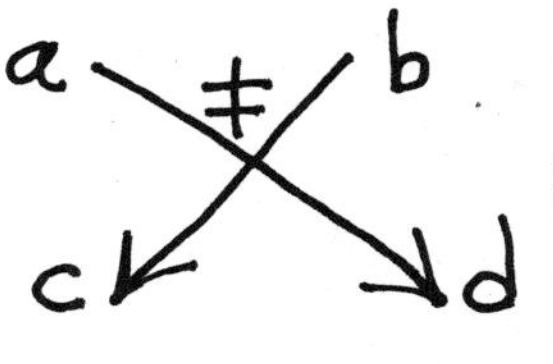

$$= \begin{cases} 1 & \text{if } a=d, b=c, \\ & \quad a \neq b. \\ 0 & \text{otherwise.} \end{cases}$$

We wish to create solutions R and R^{-1} to QYBE by taking linear combinations of these four tensors so that R and R^{-1} are inverses, and they satisfy the exchange identity

$$\mathbf{R} - \mathbf{R}^{-1} = \mathbf{zI}$$

Let R and R^{-1} be given in the form

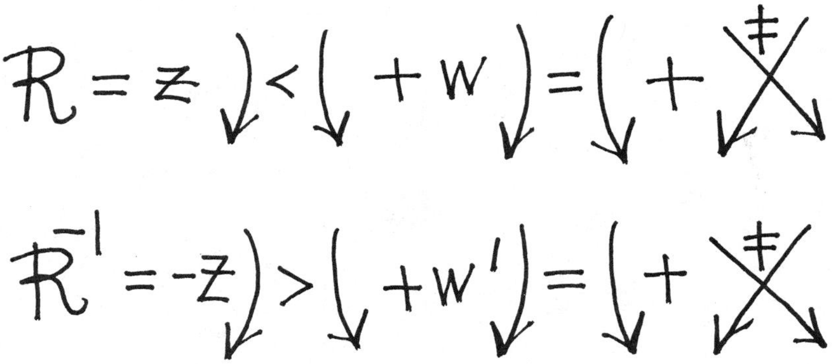

with $z = w - w'$.

Remark. It is convenient to use the crossing to which R or R^{-1} will correspond in the state model as a symbol for that matrix. I will also use the notation for vertex weights that we developed in the last section. Thus

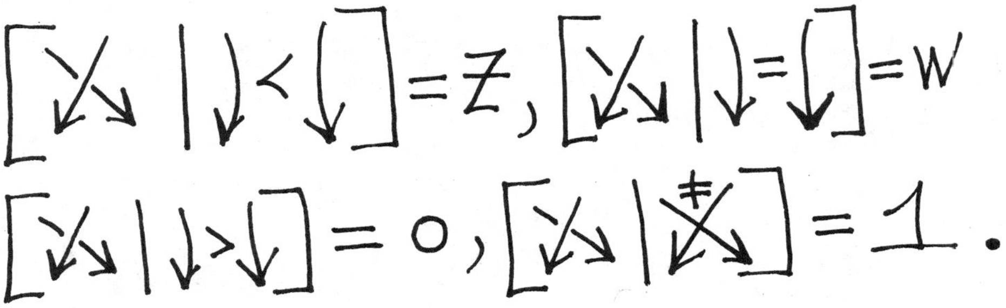

Note also that the equation

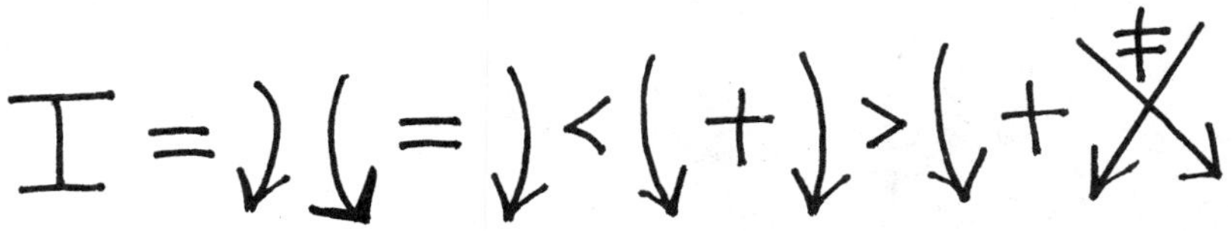

is a tautology (since two labelled lines have their indices in the one of the relations <, >, =). Therefore the exchange identity is automatically satisfied by these choices, and we need to determine conditions on z,w,w' for R and R to satisfy the QYBE.

Theorem 7.1. *Let the index set I be any subset of the real numbers. Let* $w'=(1/w)$. *Then R and* R^{-1} *as given above are each solutions to QYBE and they are inverses of each other.*

Proof. Consider a fixed triangle configuration with fixed spins on the input and output lines. More than one configuration of local states may correspond to a given choice of inputs and outputs. To fix notation, let T denote the triangle

and T' denote the triangle

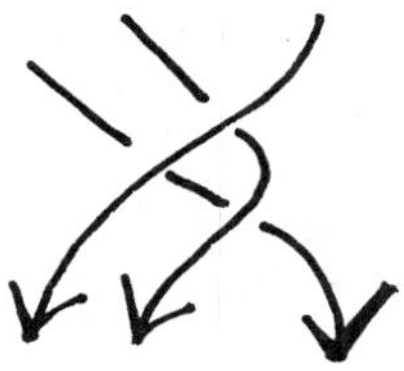

Thus

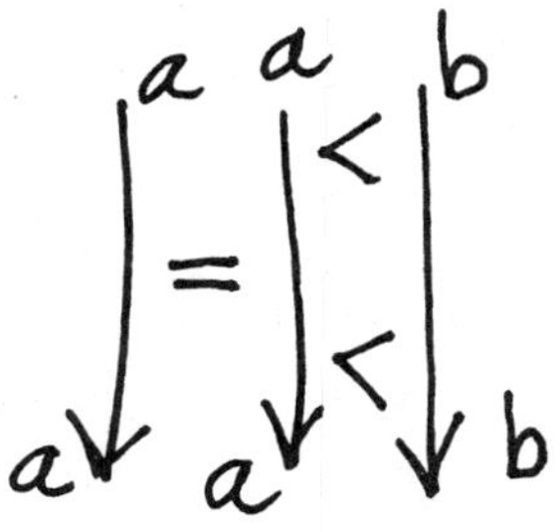

is a state of T' and

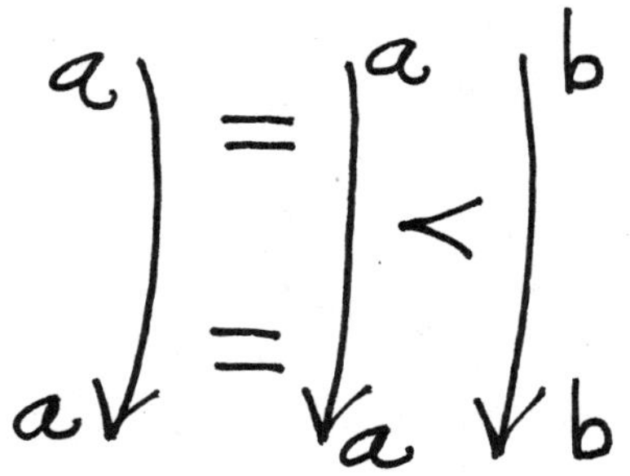

is a state of T for the input/output configuration

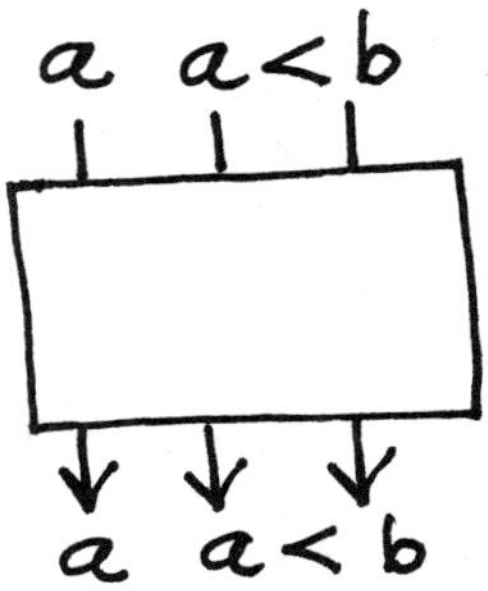

To prove QYBE it suffices to check the various cases of input and output configurations for T and T'. We check a key case and leave the rest for the reader:

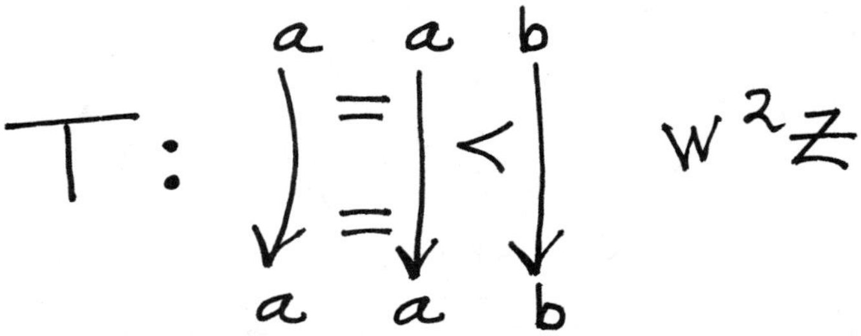

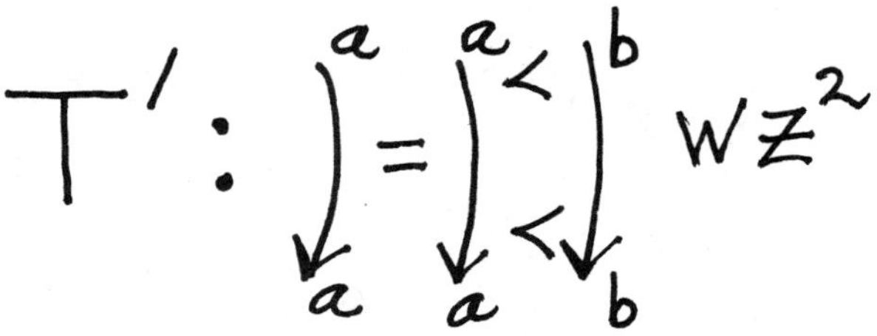

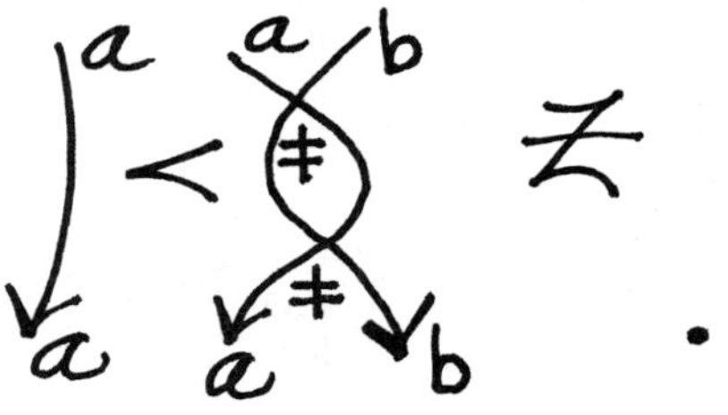

Since the contributions are the products of the vertex weights via the matrix

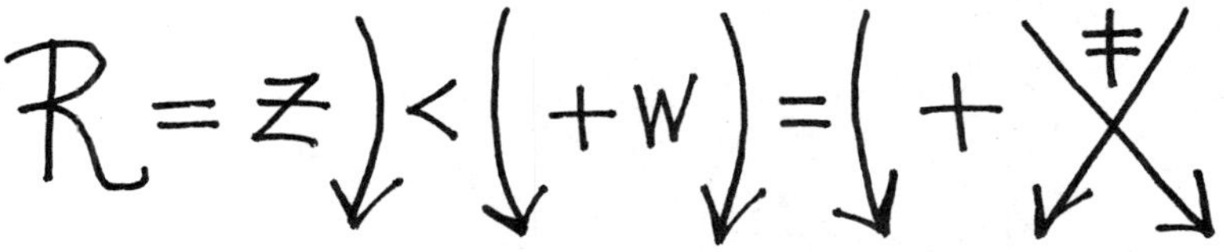

we see that invariance under the triangle move demands that

$$wz^2 + z = w^2z.$$

Since $z = w - w'$ and $w' = (1/w)$ by hypothesis, this case of invariance is verified.

Leaving the remaining cases to the reader, this completes the proof that R satisfies the QYBE. Similar work applies to R, and it is easy to see that R and R^{-1} are inverses of one another.

The Oriented State Model

Let K be an oriented link diagram. Let R and R^{-1} be the solutions to QYBE described above with index set I. Let U be the underlying locally 4-valent plane graph corresponding to K (K's universe).

Definition 7.2. A *state* S of K is a labelling of the edges of U with elements (spins) from the index set I so that spin is *strongly preserved* at each vertex. (This means that the set of the spins going into the vertex equals the set of the spins going out. Going in and going out refer to the directions of the oriented edges at the vertex.)

Since we intend to use these states with respect to the diagrammatics for R and R^{-1}, I define the *splitting , L(S), of a state S* to be the labelled universe obtained from S by splicing each vertex where the spins do not cross-over as shown below:

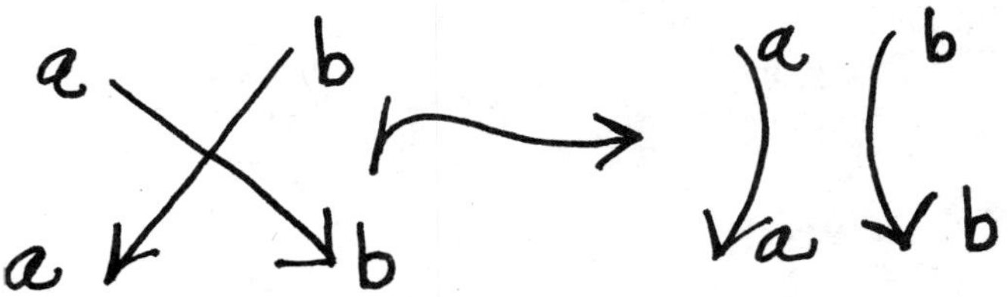

Thus L(S) is a collection of simple closed curves in the plane each labelled by a single spin from the index set. The local sites in L(S) correspond to the different conditions listed by R and R^{-1}. For example

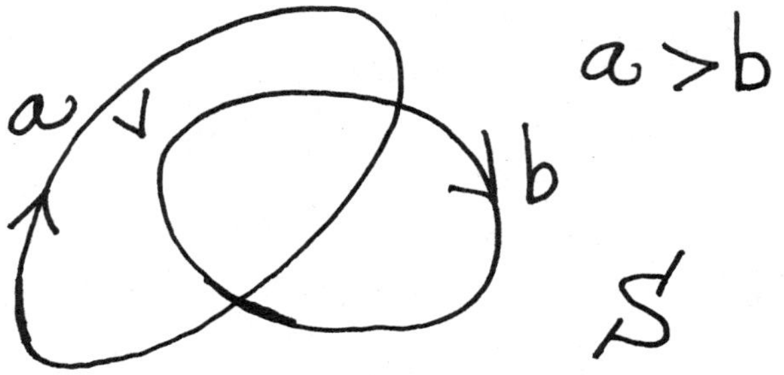

This discussion shows that an alternative definition of a state S is that it is a labelled universe , L(U) , obtained from the given universe U by splicing some of the vertices so that there are no self-crossings in L(U), and each component of L(U) is labelled with constant spin. The local decorations that we have used for the symbolic expansions of the matrices R and R^{-1} can be affixed to the sites of the state to indicate spin relationships.

Definition 7.3. The *norm of a state S* is the summation

$$||S|| = \sum_{C \varepsilon \{\text{loops in } L(S)\}} \text{label}(C)\text{rot}(C)$$

where label(C) denotes the spin assigned to the component C of L(S), and rot(C) is its rotation number as defined in section 2.

Constructing the State Model

Using the association of R with the positive crossings and R with the negative crossings, let [K|S] denote the product of the vertex weights for a given state S. This is well-defined since the matrices R and R associate unique scalars to a given choice of input and output spins.

Let λ be a new algebraic variable , commuting with w and z. The *state summation [K]* is defined by the formula

$$[K] = \sum_S [K|S] \lambda^{||S||}$$

where the *norm of S* , $||S||$, is as defined in Definition 6.3. The norm of a state is a generalization of the loop count used for the bracket polynomial. Section 8 will show a generalized version of this norm. Here we formulate the states to emphasize the loop structure, and the way in which this model grows as a natural generalization of the bracket.

In order to obtain an invariant of regular isotopy, we shall need to specialize the variable λ. However, the following result holds for the model as it stands:

Proposition 7.4. *The two-variable state summation* $[K]$ *is invariant under Reidemeister moves IIIA,IIIB and IIA. The model satisfies the exchange identity*

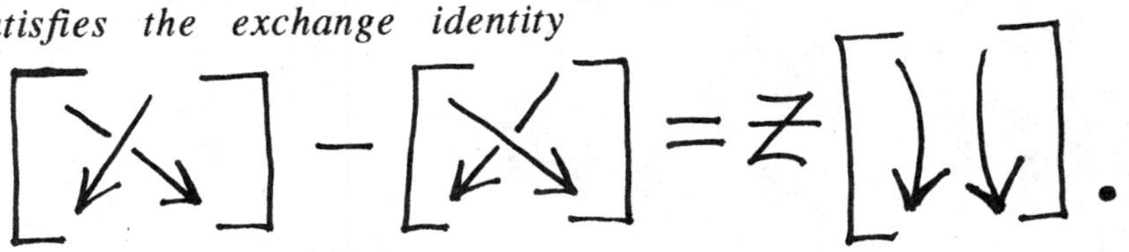

Proof. Observe , by referring to the proof of Theorem 7.1, that the norm of a state S remains invariant under all the local replacements involved in listing the cases of the triangle move. For example,diagrams of the following two types have the same rotation numbers:

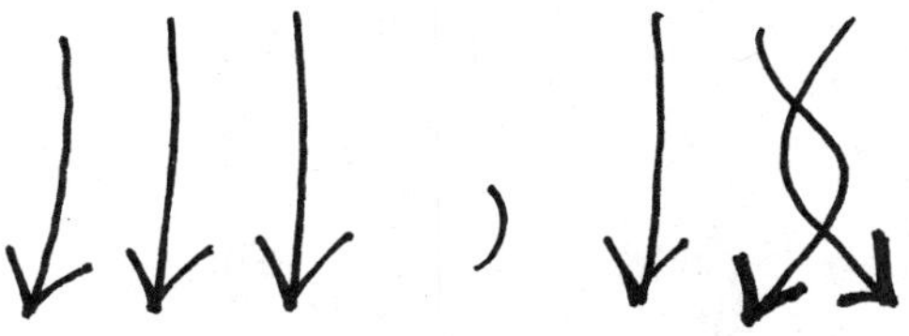

Since the state summation has the form

$$[K] = \sum_S [K|S] \lambda^{||S||}$$

this observation implies invariance under IIIA and IIIB as a direct consequence of the Quantum Yang-Baxter Equation. The invariance under IIA follows from the fact that R and R are inverse matrices. The exchange identity follows from the corresponding identity for R and R^{-1}:

$$R - R^{-1} = zI$$

This completes the proof.

Symbolic Expansion

We can write a symbolic expansion of the model [K] via

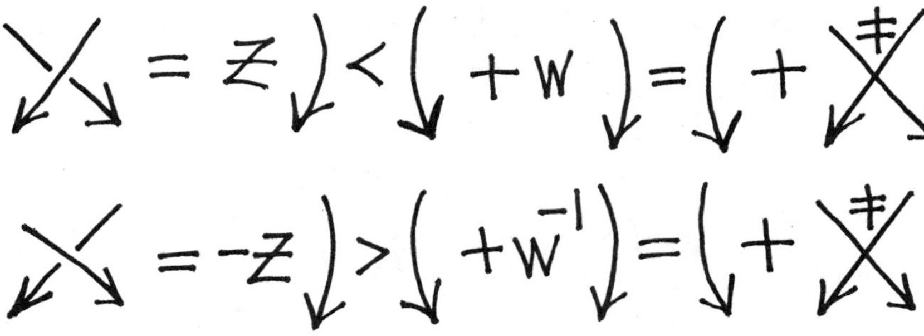

$$Z = W - W^{-1}$$

where each small diagram indicates the set of states of the larger diagram that satisfy the locally indicated relationship.

A diagrammatic proof of the invariance of the model under IIA (oriented type II move) then takes the form

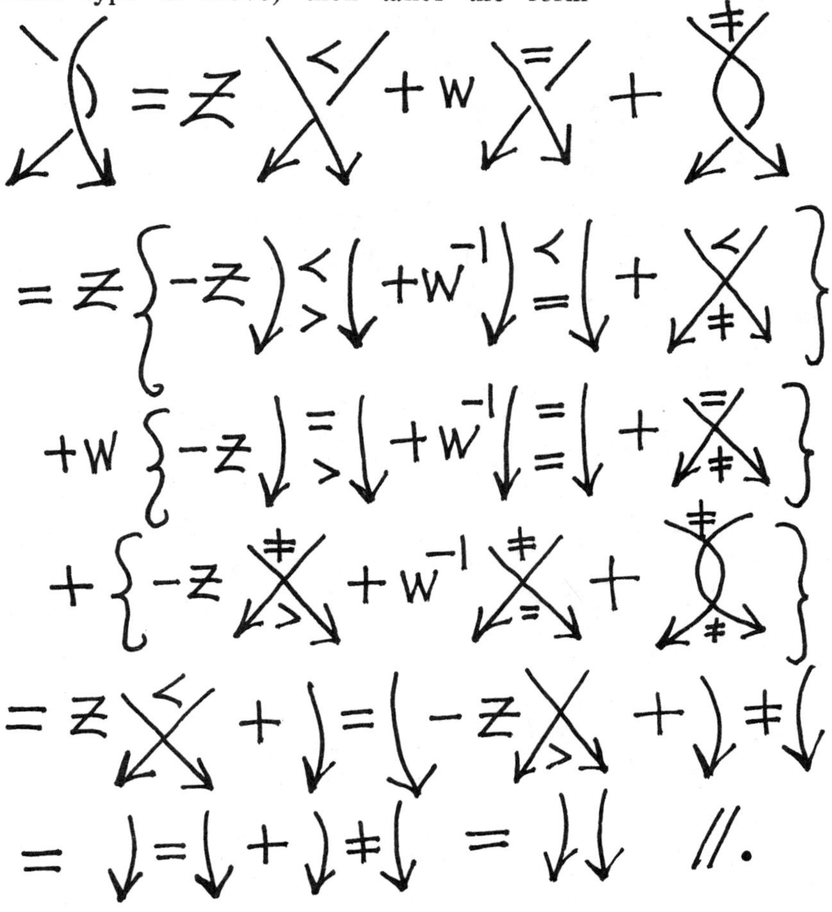

We shall use this technique to prove the

Theorem 7.5. *[K] is invariant under IIB , hence an invariant of regular isotopy, if and only if the following conditions are satisfied:*

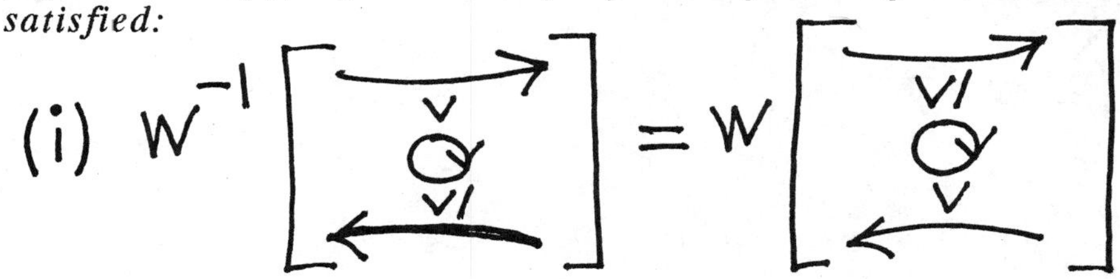

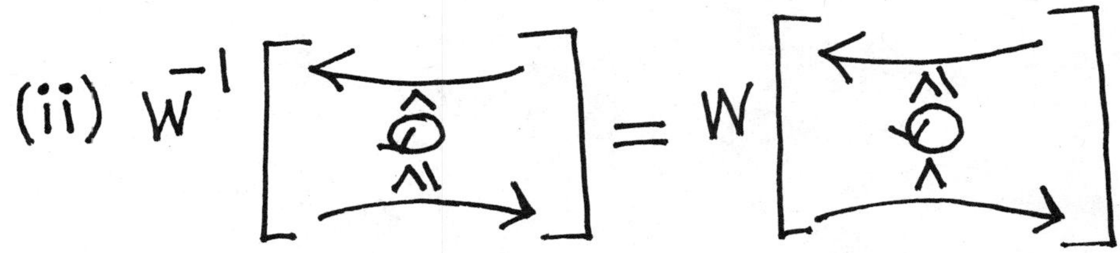

Call these the *cycle conditions* for the model.

Remark 7.6. Each of the cycle conditions corresponds to one of the orientations of the type IIB move. Each condition corresponds, potentially, to four distinct equations involving the variables w and λ , and the elements of the index set. For example, if we take the loops for the states to be in the form

then

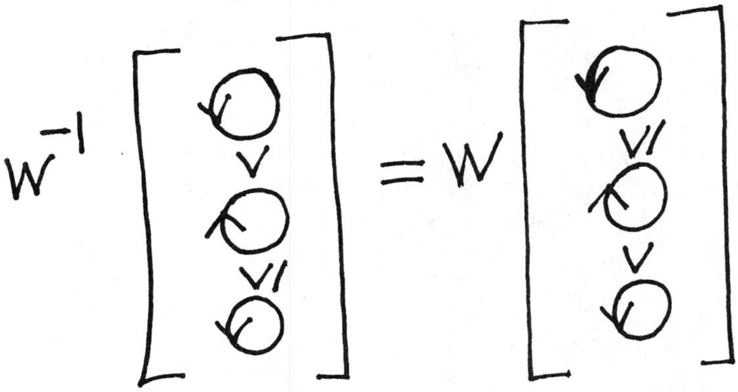

and this translates to the equation

$$w^{-1}\sum_{a>b\geq c} \lambda^{a-b+c} = w\sum_{a\geq b>c}\lambda^{a-b+c} \ ; \quad a,b,c \ \varepsilon \ I$$

Proof. The proof will be given for one choice of orientation for the type IIB move. The other case is left to the reader. $z = w - w^{-1}$.

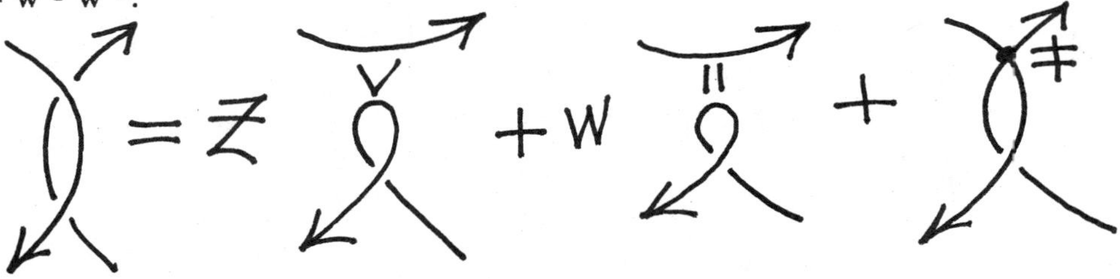

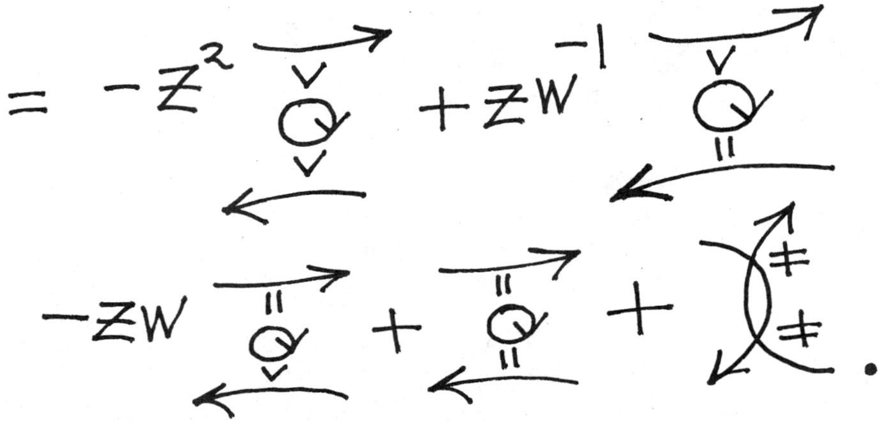

Nota bene:

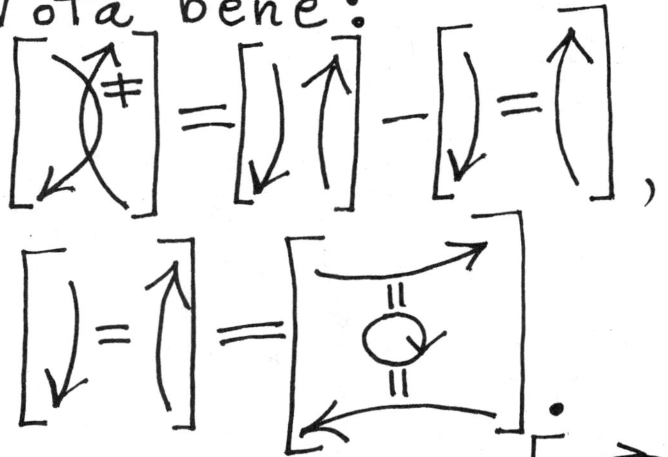

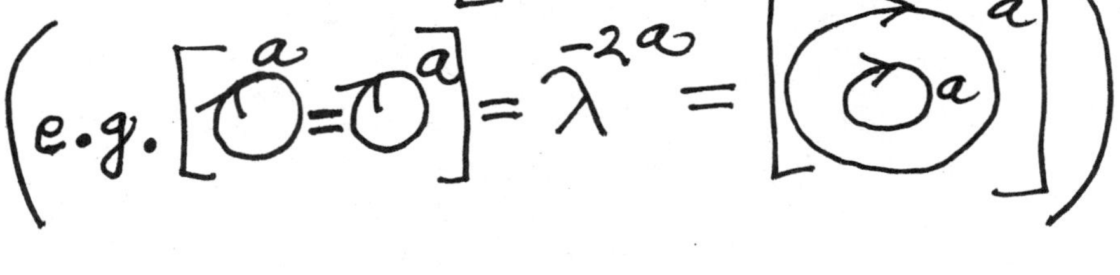

Thus we have proved the identity

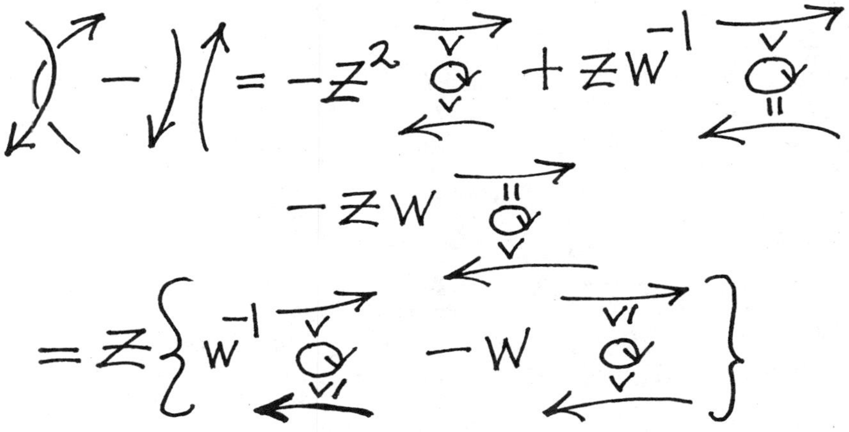

The right hand side of this equation must vanish in order for our model to be an invariant of regular isotopy. Thus we finally require:

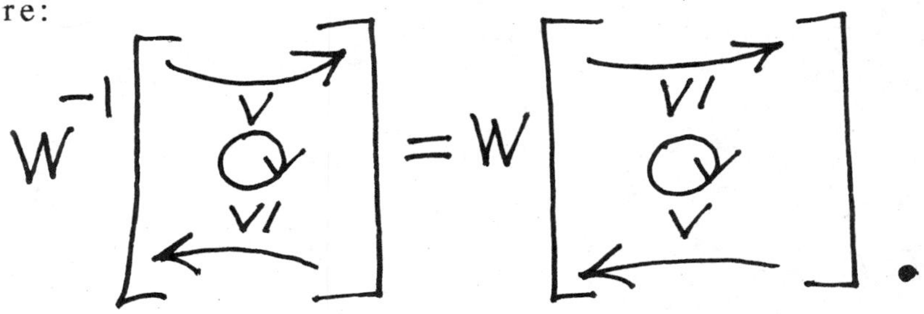

This completes the proof.

If we ask the models to behave well on type I moves, then everything simplifies:

Definition 7.7. Call the model [K] *simple* if there are scalars (elements of the base ring generated by w, w^{-1}, λ) such that

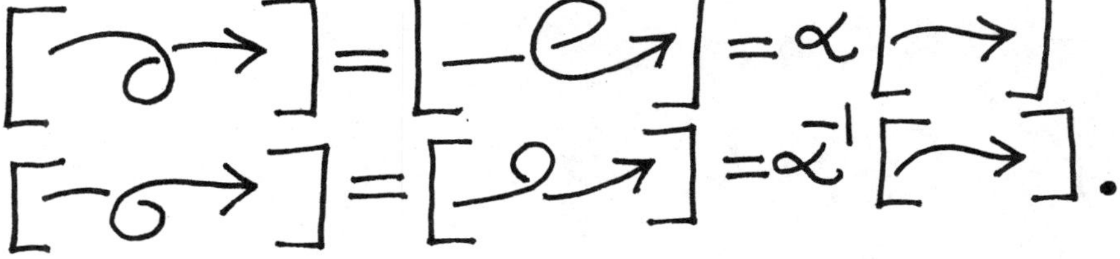

Proposition 7.8. *If [K] is simple, and the index set is of the form*

$$I = \{N_0 < N_1 < N_2 < \ldots < N_m\},$$

then all gaps $N_{(k+1)} - N_k$ *between successive elements of the index set are equal, and the lowest and highest elements are negatives of one another. In fact it is sufficient to take the index set to have the form*

$$I_n = \{-n\ ,\ -n+2,\ -n+4,\ \ldots\ ,\ n-2\ ,\ n\},$$

and in this case $\lambda = w$ *solves the cycle condition. This gives an infinite set of specializations of the Homfly polynomial that in normalized form satisfy the axioms*

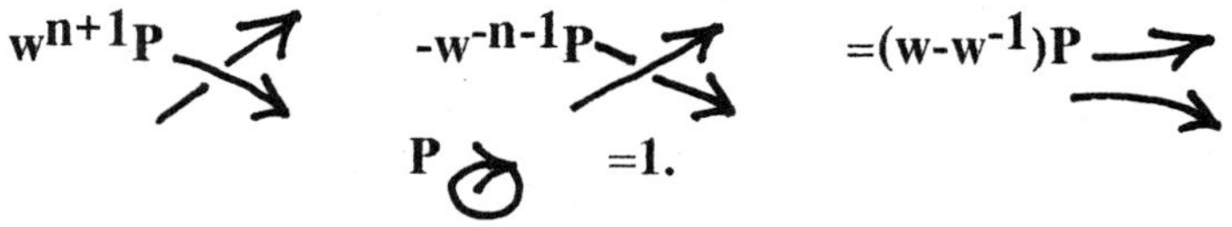

K ambient isotopic to K' implies $P_K = P_{K'}$.

Proof.

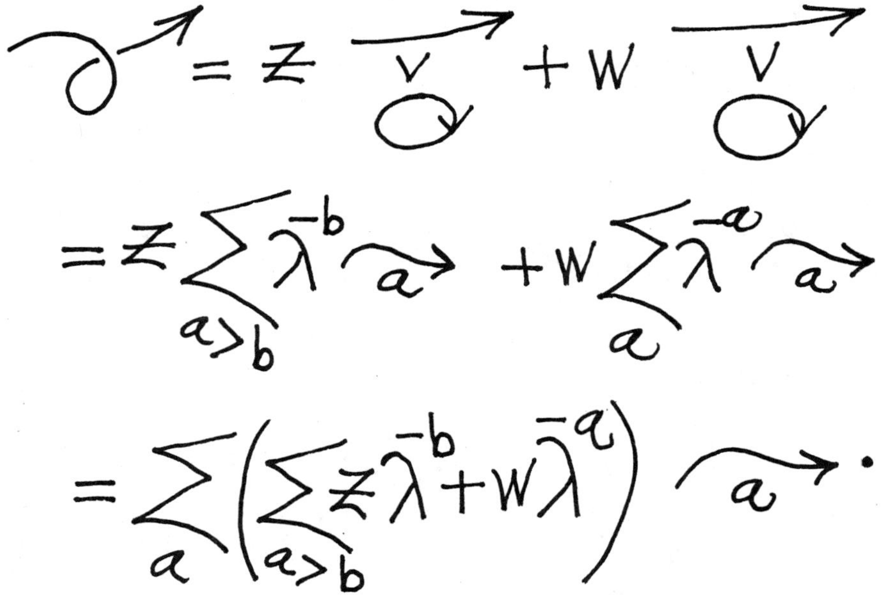

In order for the model to be simple we need that all of these coefficients are equal:

$$a = w\lambda^{-N0}$$

$$= (w-w^{-1})\lambda^{-N0} + w\lambda^{-N1}$$

$$= \ldots$$

$$= (w-w^{-1})\sum_{i=1}^{m-1} \lambda^{-Ni} + w\lambda^{-Nm}.$$

Thus

$$w\lambda^{-N0} = w\lambda^{-N0} - w^{-1}\lambda^{-N0} + w\lambda^{-N1}$$

$$\text{Hence,} \quad w^{-1}\lambda^{-N0} = w\lambda^{-N1}$$

$$\text{Hence,} \quad w^2 = \lambda^{N1-N0}$$

$$\text{and,} \quad w^2 = \lambda^{N(k+1)-Nk}, \quad k=0,1,\ldots m-1.$$

The rest of the theorem follows from this.

The verification that $\lambda = w$ satisfies the cycle condition in the case of the index set of gap two is the content of the following

Lemma 7.9. *Let* $I = \{-m,-m+2,\ldots,m-2,m\}$ *so that the summations below refer to this index set. Then the cycle conditions of Theorem 7.5 are satisfied by* $\lambda = w$.

Proof of Lemma. We must prove that

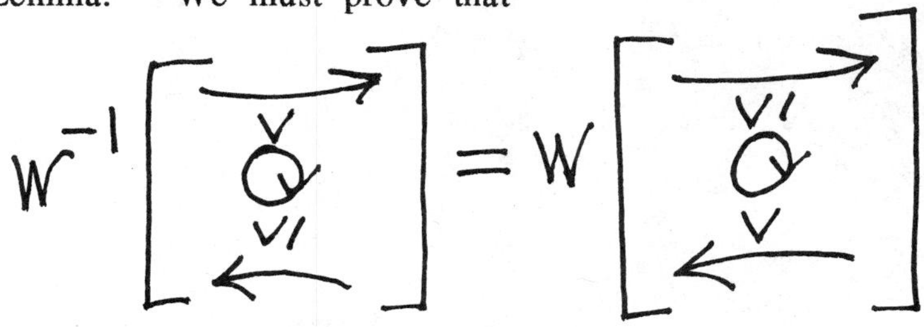

and

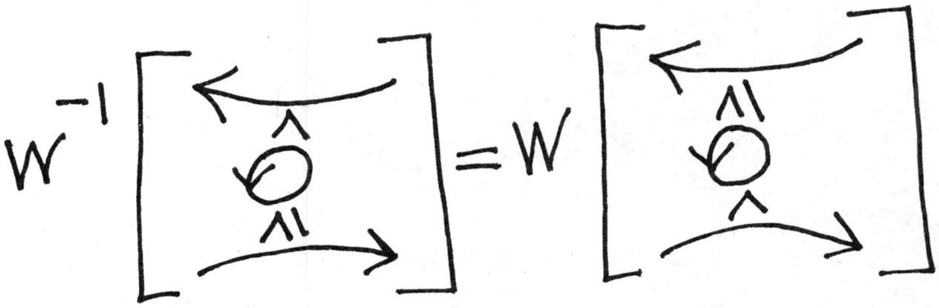

However, under the hypotheses of the Lemma, we have

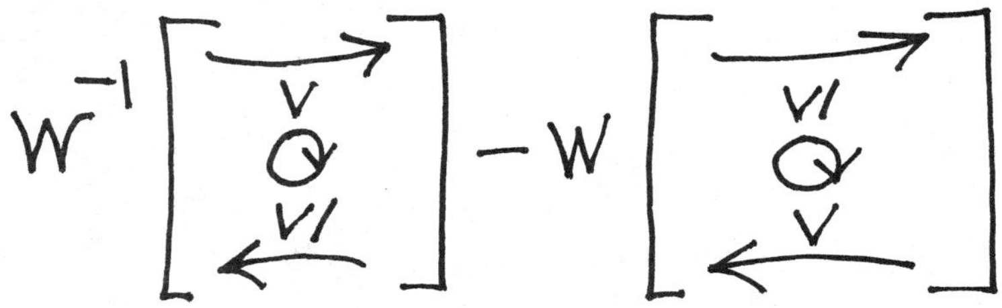

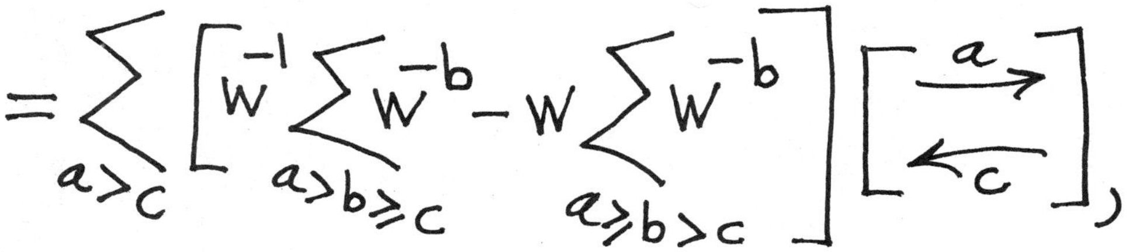

and each term

$$w^{-1}\sum_{a>b\geq c} w^{-b} - w\sum_{a\geq b>c} w^{-b}$$

vanishes identically, since the index set is equally spaced with intervals of size two. The second case goes through in the same way. This completes the proof.

Acknowledgement. We thank Mark Hennings for substantially simplifying the proof of this lemma, and for useful comments about the structure of this model.

Remark. Note that with the index set in the form

$$I = \{-m,-m+2,\ldots,m-2,m\}$$

we have that

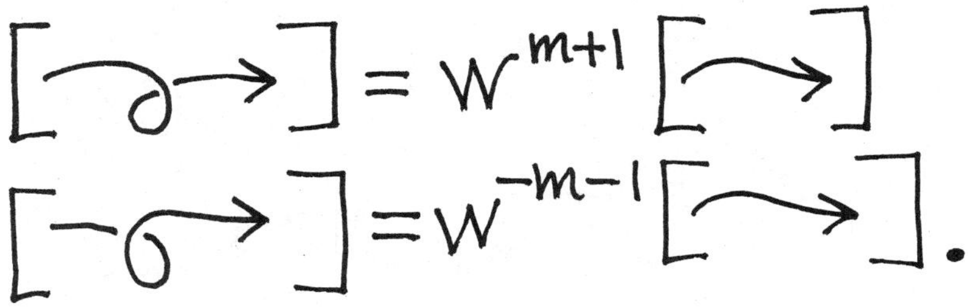

Thus, if we define $P_K(w) = (w^{m+1})^{-w(K)}[K]/[O]$, then P_K is an ambient isotopy invariant of oriented links satisfying the equations

$w^{m+1}P$  $- \; w^{-m-1}P$  $= (w-w^{-1})P$ 

$$P_O = 1.$$

In particular, if the index set has two elements $\{-1,+1\}$, then P_K is the original Jones polynomial.

Remark. The states for these models can be regarded as labellings of diagrammatic states obtained from a given link diagram K by either smoothing a crossing or by projecting it into the plane. In the resulting collection of plane curves it is required that each oriented curve is free of self-crossings. (Each curve is a curve of constant spin, and spins that cross have different values in the model.) The resulting collection of diagrammatic states for a given link diagram is remarkably sparse. For example, there are only forty such states fopr the knot 9_{42} (as opposed to 512 diagrammatic states for the bracket). Figure 12 indicates the collection of diagrammatic states for the knot 6_1. There are three diagrammatic states. In the case of the one-variable Jones polynomial (index set of order two) only one of these states contributes to the polynomial.

It is very interesting to note how the demand for simplicity in this model causes the spectrum (that is, the index set) to become equally spaced. This is a combinatorial analogue of the way the angular momentum spectrum becomes equally spaced in the standard quantum mechanics of spin.

The Knot 6_1 and Three Diagrammatic States
Figure 12

8. Yang-Baxter Models for Skein Polynomials - Unoriented Models

Recall from section 3 that the Kauffman polynomial (Dubrovnik version) has exchange identities

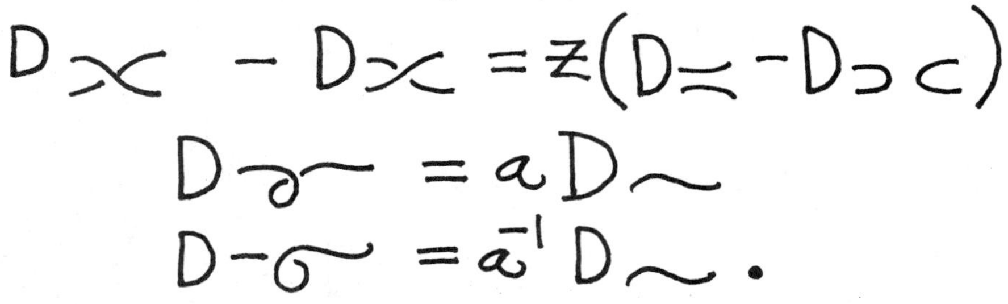

I will show in this section that a very natural variation of the state model of section 6 gives an infinite series of specializations of the Dubrovnik polynomial. (Compare [67].)

In this development, the solution M to the Quantum Yang-Baxter Equation arises from considerations involving link diagrams. This same tensor M comes from the quantum group associated with SO(n). The way it dovetails into the link-theoretic context raises questions about the relationship of the geometry of link diagrams and the theory of quantum groups. I will make some of these questions more precise at the end of this section.

The heuristic that we use in this section is the following: Consider creating a state model for the Dubrovnik polynomial that is as close as possible to the model of the previous section. Such a model could be obtained by allowing loop states formed by smoothing each crossing in two possible ways , or by projecting it:

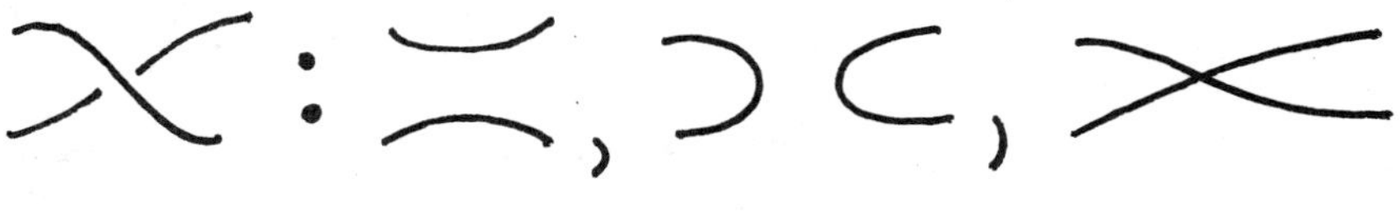

After forming such a loop state (without self-crossings as before) it will be labelled by

1. choosing an orientation for each component
2. labelling each component from the index set.

Thus we will obtain an invariant that is independent of orientation by summing over all possible orientations of the states.

One further rule simplifies matters:

3. If the orientation of a component of a loop state is reversed, then its label is negated.

With this rule , we see that the local state configurations at a crossing can all be put in the forms:

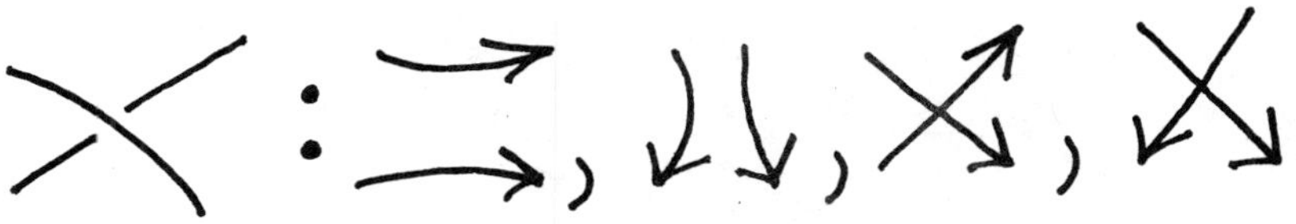

These are horizontal and vertical versions of the local configurations for the oriented case. It is suggestive to have the model satisfy an expansion of the form:

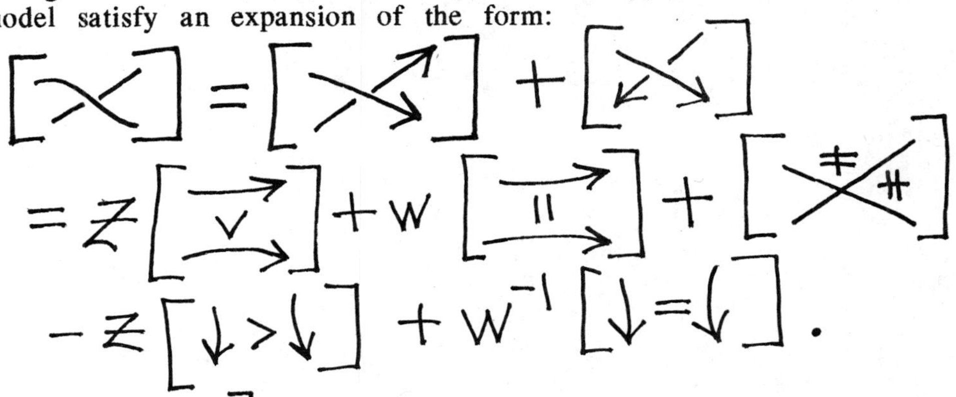

where 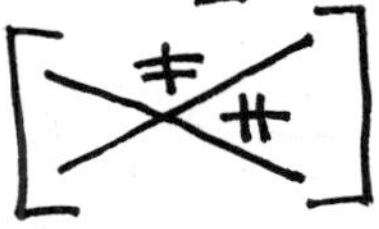 denotes summation over states where one has top inputs a and b with $a \neq b$ *and* $a \neq -b$: 

Note that this expansion gives, formally, the correct exchange identity for the Dubrovnik polynomial.

Following the parallel with the previous section we take this expansion as defining the local vertex weights, and use the global formula:

$$[K] = \Sigma_S \ [K|S] \ w^{||S||}$$

with the vertex weights specified as shown below: $(z=w-w^{-1})$

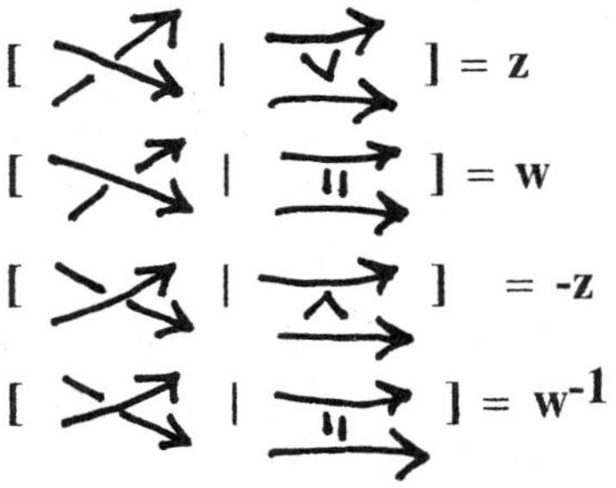

if $a \neq b$ and $a \neq -b$

if $a \neq b$ and $a \neq -b$.

Here $[K|S]$ denotes the product of the vertex weights (over all crossings) for a given state S.

The norm of a state is defined just as in the previous section. The question now becomes whether there is an index set for which this model is a regular isotopy invariant.

Having come upon this model by such a backwards heuristic, we must ask -- what is the scattering matrix? At first glance it may seem that the scattering matrix is

by reading directly from the local vertex weights.

However, a state with a local configuration in the form

receives a different rotation number from an otherwise identical state with local configuration

in those cases where both splices can occur. Thus *an abstract check of triangle invariance for the model as a whole requires a compensation for the rotation number.* The example below illustrates this phenomena

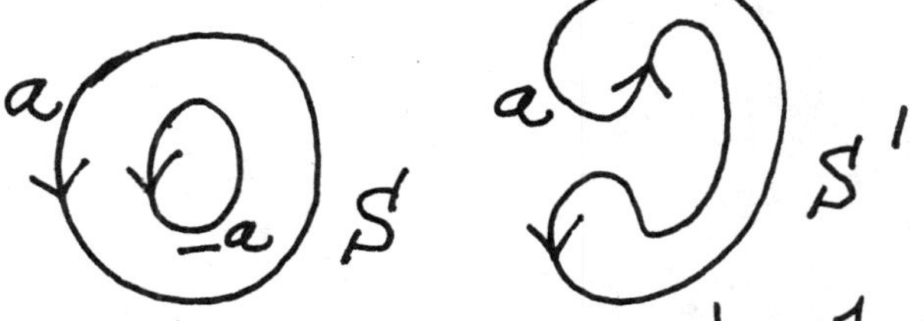

$$\text{rot} = 0 \qquad\qquad \text{rot} = 1$$

$$\|S\| = 1 \qquad\qquad \|S'\| = (a-(-a))/2 = a,$$

and we conclude that the general compensation is given by a power of w as indicated in the diagram below:

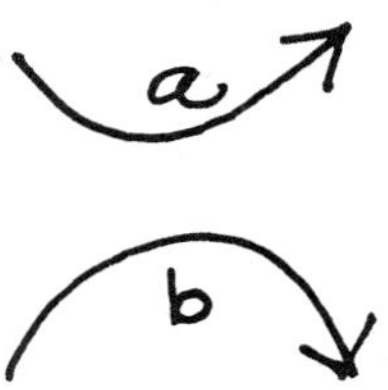

$$\Longleftrightarrow W^{\frac{a-b}{2}}$$

Here is a second example:

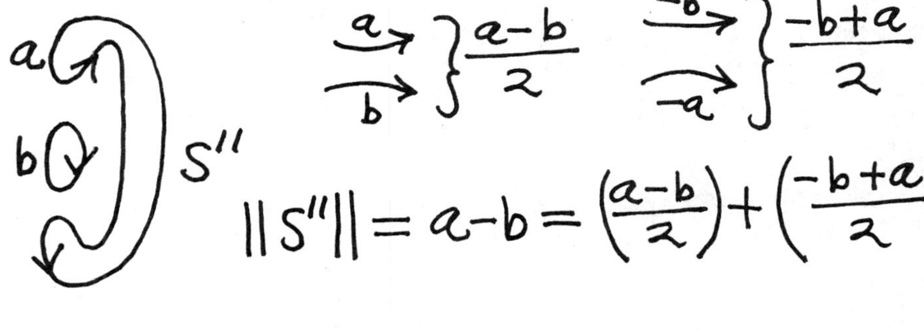

$$\|S''\| = a-b = \left(\frac{a-b}{2}\right) + \left(\frac{-b+a}{2}\right).$$

It follows that *triangle invariance of this model is equivalent to the following matrix being a solution to the QYBE:*

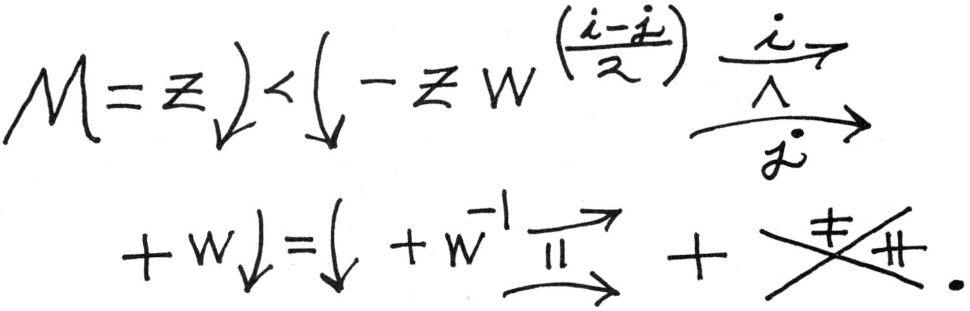

Here it is tacitly assumed that in the crossover the indices on different lines are of different absolute values.

There is a further subtlety in this model. In order to expose this special feature, I shall define two types of index set that actually work for the model , and the rules for using them.

Definition 8.1. For n odd, let $\mathbf{I_n}$ denote the index set

$$\{-n,-n+2,-n+4,\ldots,-1,0,1,3,\ldots,n-2,n\}.$$

For n even , let $\mathbf{I_n}$ denote the index set

$$\{-n,-n+2,\ldots,-2,0,\hat{0},2,4,6,\ldots,n-2,n\}.$$

Note that for odd n the index 0 is special in that it is spaced by one unit from its neighbors, while all the other indices have spacing equal to two. For even n , there are two "zero" indices. They are distinct indices, but will behave numerically as zeroes.

Zero Rules.
Here are the rules for handling these zeroes.

1. If n is odd, then SHIFT the occurrence of

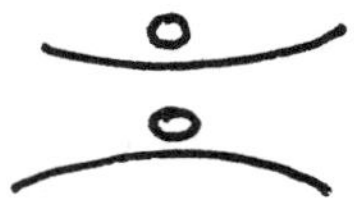

to an occurrence of

and assign it a vertex weight of 1.

This means that the R matrix now takes the form

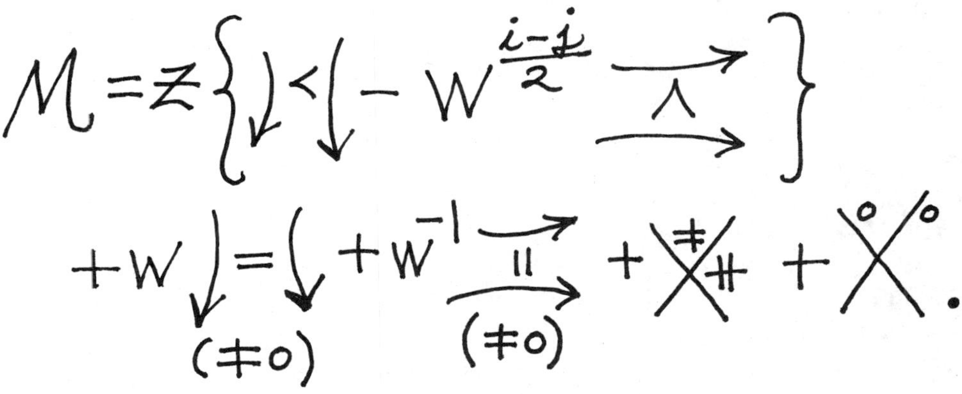

2. For n even, treat the extra zero just as another copy of a zero index.

In all cases the value of a zero index is zero for norm computations.

States. States are labelled oriented partial smoothings of the universe of K, so that no component has any self-crossings (except in the case of the SHIFT for the zero index with n odd.) The norm is defined as in the previous section. Note that we assume that the "norm variable" is w.

The first reason for these special conventions about the index set presents itself when we try to verify that the model is multiplicative under the type I Reidemeister move. The extra configuration

creates a difference in the summation. The extra zeroes correct for this difference. It then happens that these corrections on the index set are also needed in order that the matrix M satisfy the Yang-Baxter Equation.

Proposition 8.2. *Let [K] be defined for unoriented diagrams as described above. If the matrix M above is a solution to the QYBE, then [K] is invariant under the type IIIA move.*

Proof. As observed above, the extra rotational term on the matrix serves to insure that in summing over all states contributing to a triangle move, identical rotational contributions can be factored from each term. This leaves the indicated factors on the matrix terms. If the matrix, so configured, satisfies the QYBE then the model will be triangle invariant.

In fact the matrix M does satisfy QYBE for the special index sets.

Proposition 8.3. *Let the index set be taken in the form given in Definition 8.1 with the surrounding conventions for the zeroes. Then M is a solution to QYBE.*

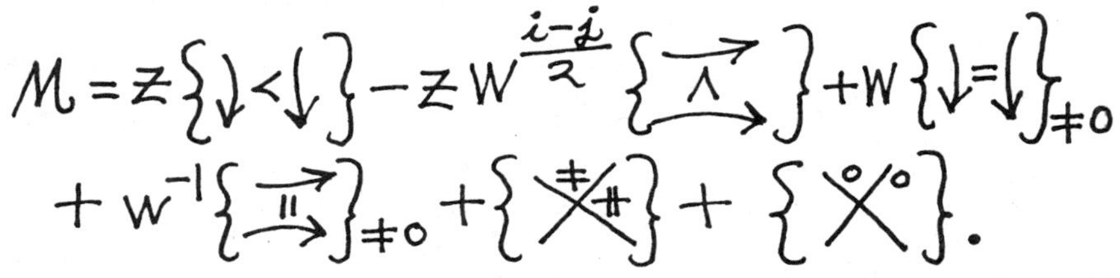

Proof. See [39] pp. 229-231. //

Proposition 8.4. *With the index set chosen as in the previous proposition, [K] is an invariant of regular isotopy. These models are multiplicative under the type I move. That is, there are scalars a and a^{-1} such that*

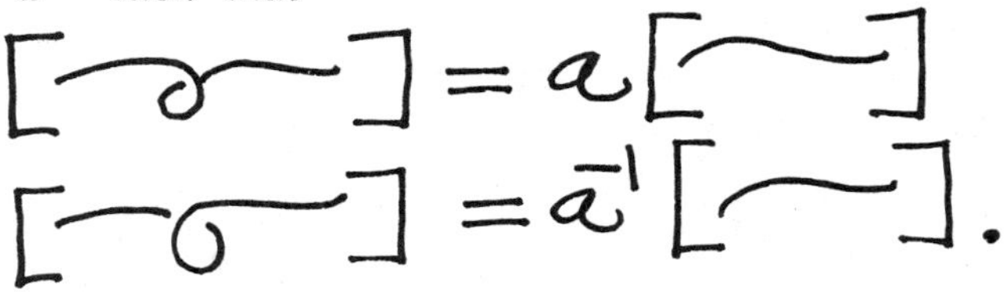

Hence this gives a set of state models for the Kauffman polynomial.

Proof. The verification is straightforward and we omit it.//

Remark. What is most remarkable about this model is how it and its scattering matrix arise in the context of link diagrams. The context of knot theory and these state models helps explain the reason for the rotational factors, but the special zeroes remain more mysterious. In the context of quantum groups, the models with index set I_n correspond to the group SO(n). The relationship of this approach with the representation theory needs deeper study. Compare particularly this approach with the work of Reshetikhin [59].

9. Piecewise Linear Models

In this section I show how the models of section 7 and 8 are special cases of a more general model that is motivated by particle scattering.

We have already interpreted the Yang-Baxter Equation as stating an identity about the scattering matrices for idealized two-particle interactions. But so far, the only properties that these particles had was a discrete quantity that we called "spin". The spin came from a discrete ordered index set.

To be more realistic, one can include a measure of relative momentum in the form of a rapidity parameter theta , θ . That is, the scattering matrix is a function of an angular parameter θ expressed in radians:

$$S = S^{ab}{}_{cd}(\theta).$$

In this section we shall consider models whose local vertex weights are expressed by such an S-matrix where S has the form

$$S = R^{ab}{}_{cd}\ \lambda^{(d-a)\theta/2\pi}$$

and the angle is represented diagrammatically as the angle between the line segments in the crossing. See Figure 13.

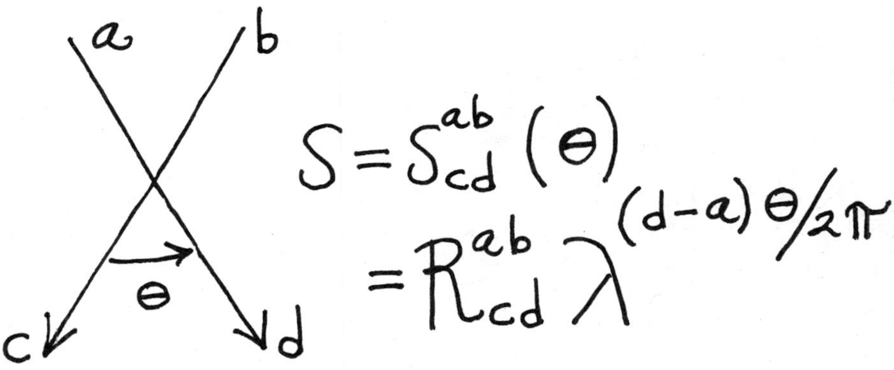

Figure 13

Here we assume that the R-matrix is zero unless $a+b=c+d$. Thus total spin is preserved in the interaction. More will be said about this angular factor as the discussion continues.

The Rapidity

A word of background about this rapidity parameter is in order (See [7] or [22]). The underlying assumption is that the particles are interacting in the context of special relativity. This means that the momenta of two interacting particles can be shifted by a Lorentz Transformation and the resulting scattering amplitude must remain invariant. Now , in the relativistic context (with light speed equal to unity) we have the fundamental relationship among energy (E), momentum (p) and mass (m):

$$E^2 - p^2 = m^2$$

(Note that this is in 1+1 dimensional spacetime.) Letting the mass be unity, we have

$$E^2 - p^2 = 1$$

and can represent $E=\cosh(\theta)$, $p=\sinh(\theta)$.

The Lorentz transformation shifts theta by a fixed amount. In light cone coordinates the Energy-Momentum has the form

$$[\ e^{\theta}\ ,\ e^{-\theta}\ ]\ ,$$

and this is transformed via

$$[A,B] \text{ -----> } [\ Ae^{\phi},\ Be^{-\phi}\ ]$$

under a Lorentz transformation. Consequently, if the two particles have respective **rapidities** (the theta parameter) θ_1 and θ_2, then the difference $\theta_1 - \theta_2$ is Lorentz invariant. The S-matrix must be a function of this difference of rapidities.

Now consider the diagram in Figure 14. Since momentum is conserved in the interactions, we see that for the oriented triangle interaction, the middle term has a rapidity difference that is the sum of the rapidity differences of the other two terms. But the angles between the corresponding lines satisfy just this relation - by a Theorem well-known to Euclid.

Thus we can identify the rapidity differences with the angles between the interaction lines and obtain the correct angular relationships in the geometry of the diagrams.

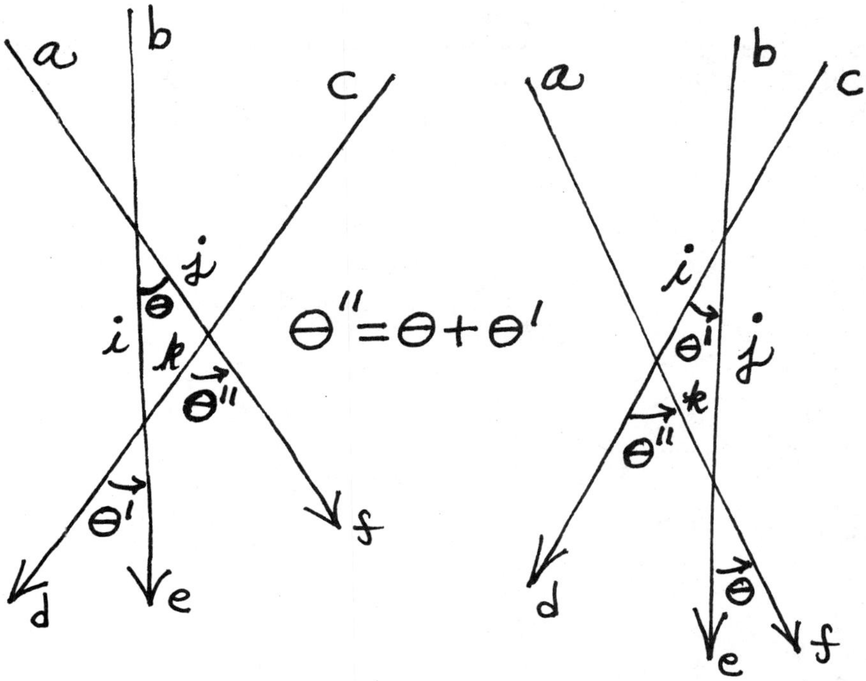

$$S^{ab}{}_{ij}(\theta)S^{jc}{}_{kf}(\theta+\theta')S^{ik}{}_{de}(q')=S^{bc}{}_{ij}(\theta')S^{ai}{}_{dk}(\theta+\theta')S^{kj}{}_{ef}(\theta)$$

Quantum Yang-Baxter Equation with Rapidity
Figure 14

Piecewise Linearity and Spin States

The link diagrams are now taken to be piecewise linear. This means that the underlying plane graph of the diagram has linear segments for all of its edges and that the pairs of segments that make crosslines form straight angles. See Figure 15.

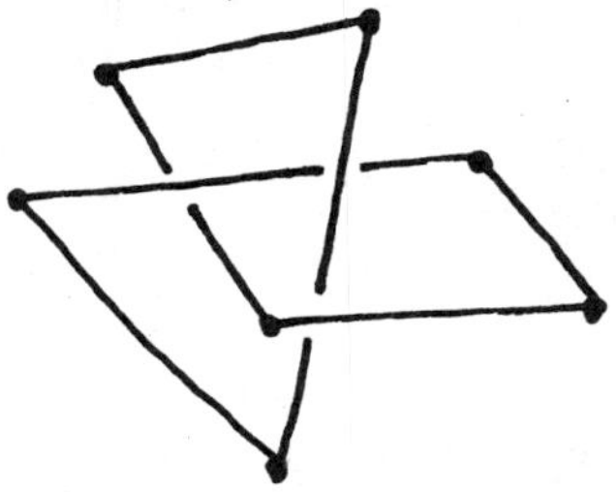

Piecewise Linear Diagram
Figure 15

A piecewise linear link diagram has crossing vertices and extra vertices of valence two. By assigning vertex weights to the crossings and to the valence-two vertices, we shall form a state model. The vertex weights are as shown in Figure16.

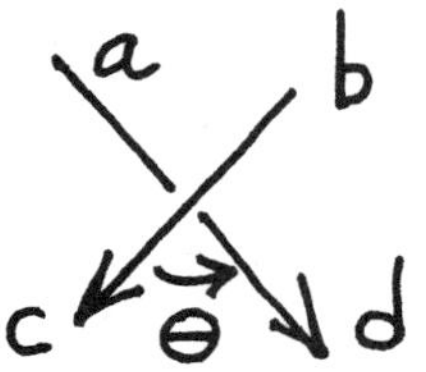

<-----> $S^{ab}_{cd}(\theta) = R^{ab}_{cd}\ \lambda^{(d-a)\theta/2\pi}$

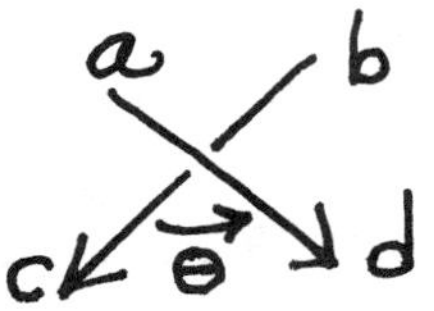

<----->$(S^{-1})^{ab}_{cd}(\theta)=(R^{-1})^{ab}_{cd}\lambda^{(d-a)\theta/2\pi}$

<-----> $\lambda^{a\theta/2\pi}$

Vertex Weights
Figure 16

We associate the matrix S to a positive crossing, and the matrix S^{-1} to a negative crossing. Here it is assumed that S^{-1} is the inverse to S in the sense that

$$S(\theta)S^{-1}(-\theta) = I.$$

Note that the angular dependence involves only the underlying plane graph of the diagram. In the case of a two-valent vertex, the weight is lambda -- raised to the power the average of the incoming and outgoing spins, multiplied by the angle divided by 2π. The angle is measured in the direction of turn from the ingoing to the outgoing lines.

In most cases, we shall assume that the (spin) states of the network are *spin-preserving.* This means that at a crossing the sum of the in-spins is equal to the sum of the out-spins, and ;that at a two-vertex the in and out spins are equal. As in sections 6 and 7, a *state* of a link diagram K is an assignment of spins to the edges of its underlying plane graph.

Let $< K \mid S >$ denote the product of all the vertex weights for a given state . Let $<K>$ denote the sum of these products over all spin-preserving states of the diagram K:

$$\langle K \rangle = \sum_S \langle K | S \rangle$$

Call $\langle K \rangle$ the *amplitude* of K with respect to the given S-matrix.
I will assume that the S-matrix is *factorized*:

$$S^{ab}{}_{ij}(\theta) S^{jc}{}_{kf}(\theta+\theta') S^{ik}{}_{de}(\theta') = S^{bc}{}_{ij}(\theta') S^{ai}{}_{dk}(\theta+\theta') \quad S^{kj}{}_{ef}(\theta).$$

Here the repeated upper and lower indices are summed over the index set I. This is is the *Quantum Yang-Baxter Equation* with the rapidity parameter included. Note that the appearance of the sum of the two angles corresponds to the angles and angle sums in Figure 14

Lemma 9.1. *Given the matrix S in the form*

$$S^{ab}{}_{cd}(\theta) = R^{ab}{}_{cd} l^{(d-a)\theta/2\pi} \quad ,$$

S is factorized if and only if **R** *satisfies the QYBE without rapidity parameter. That is,* **R** *must satisfy*

$$R^{ab}{}_{ij} \; R^{jc}{}_{kf} \; R^{ik}{}_{de} = R^{bc}{}_{ij} \quad R^{ai}{}_{dk} \; R^{kj}{}_{ef}.$$

Proof. We check, using Figure 14, that the total angular contribution of the triangle depends only on the input and output spins.

$$\begin{aligned}
\Delta &= \text{angle contribution} \\
\Delta &= (j-a)q \;\; +(f-j)(q+q') \;\; +(e-i)q' \\
&= (f-a)q \;\; +(f-j)q' \;\; + \;\; (e-i)q' \\
&= (f-a)q \;\; + \;\; (c-k)q' \;\; +(k-d)q' \\
&= (f-a)q \;\; + \;\; (c-d)q'.
\end{aligned}$$

Therefore, the term

$$\lambda^{\Delta/2\pi}$$

can be factored from the Yang-Baxter condition , leaving only the angle-free equation for R. This completes the proof. //

We now rewrite the amplitude by collecting together the angular terms. Define the *angular value* of a state, at a vertex v of the diagram K, to be the exponent in the vertex weight. That is,

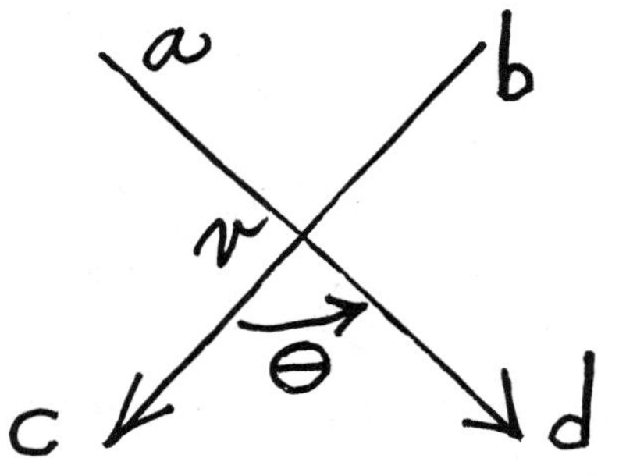

$$\sigma(v) = (d-a)/2\pi$$

Definition 9.2. Define the *norm* of a spin state, $||\sigma||$, to be the sum of the angular values at its vertices. Thus

$$||\sigma|| = \Sigma_v \ \sigma(v).$$

Given a state σ, let $<K \mid \sigma>$ denote the product of the R-matrix contributions to the product of vertex weights. Then

$$<K \mid \sigma> = [K \mid \sigma] \ \lambda^{||\sigma||},$$

and the amplitude now takes a form that is analogous to the models of sections 6 and 7:

$$<K> = \Sigma_\sigma <K \mid \sigma> = \Sigma_\sigma [K \mid \sigma] \ \lambda^{||\sigma||}.$$

Definition 9.3. A spin state of a diagram is said to be *simple* if it is spin-preserving, and the set of input spins at each crossing is equal to the set of output spins at that crossing.

Recall from section 7, that we associated to each simple spin state, a *loop state* consisting in the oriented loops of constant spin. The set of such loops is obtained by splicing or projecting each crossing according as the spins are not permuted or permuted, respectively. The *loop norm* was defined to be the sum of the rotation numbers of the loops weighted by their respective spins.

Lemma 9.4. *The norm of a simple spin-state, as defined above, is equal to the loop-norm of its corresponding loop state.*

Proof. First note that the angular contribution of a crossing can be written as the sum of two vertex contributions as shown in

Figure 17. The two vertices are obtained by splitting the crossing in oriented fashion. In the case of the simple spin state, we apply this splitting to form the loops. Thus , each loop can be regarded as supplying the sum of its own angular values. It is easy to see that the sum of angular values for a given loop is equal to the spin label of the loop multiplied by its rotation number (The sum of the exterior angles of a simple closed polygon is 2π.). This completes the proof. //

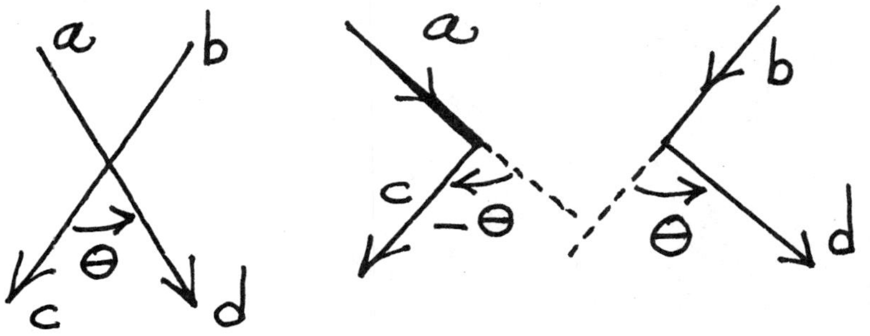

$$(d-a)\theta=((a+c)/2)(-\theta)+((b+d)/2)\theta$$
when $a+b=c+d$.

Figure 17

Remark. It follows from this lemma that the oriented model of section 6 is a special case of the piecewise linear model discussed in this section.

Reidemeister Moves. In order to discuss the invariance of the piecewise linear model under Reidemeister moves, we must briefly discuss their PL (piecewise linear) version. Representatives of these moves are shown in Figure 18. Note that the input and output lines for the type II moves are depicted as parallel, and that segments are moved in a parallel fashion in the type III move as well. All this speciality is allowed because of the presence of the (now articulated) type IV move that allows a piece of a straight segment to be replaced by the other two sides of a triangle (initiating three new vertices in the process). The type IV move includes the reverse of this procedure, and it makes explicit in the PL category what we had called equivalence of the underlying graphical structure.

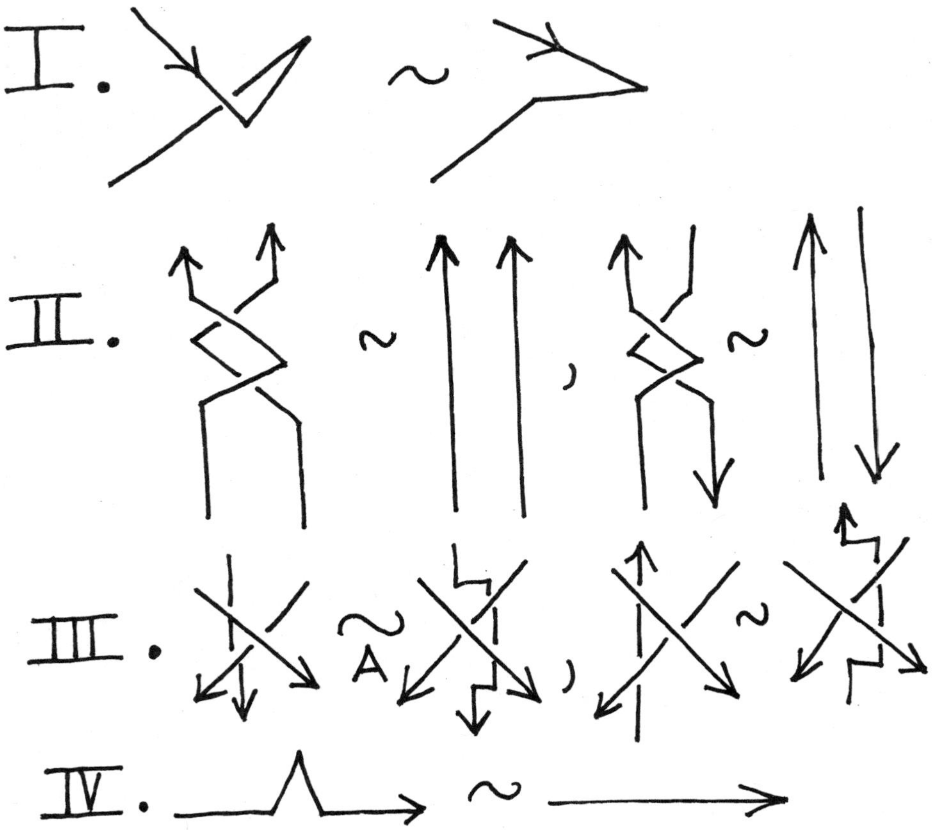

Piecewise Linear Reidemeister Moves
Figure 18

A historical note is in order. Reidemeister [58] originally worked with PL diagrams, and in fact a version of the type IV move in three dimensional space generates PL ambient isotopy of knots and links in three-space. The moves of type I,II, and III are the result of articulating the shadows of the combinatorial isotopies in three dimensional space.

Note that doing a type IV move and repeating constant spin on the new edges does not change the norm of the state.

Notation. I shall write a wavy line between two separated vertices in a spin state to indicate that they are to be regarded as one site of interaction with spin conservation as indicated below:

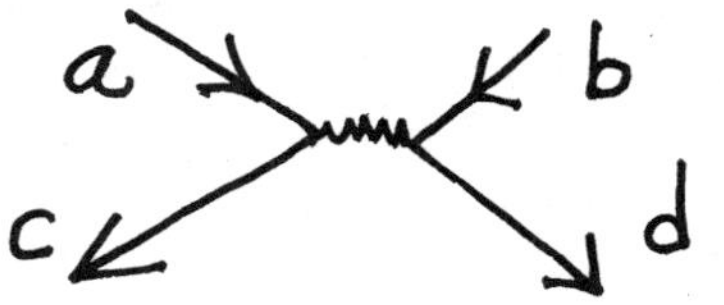

a+b = c+d.

This notation allows us to work with a generalization of spin states that makes the next Lemma easier to state. This Lemma gives the angle data for examining behaviour under the two orientations of the type II move.

Lemma 9.5. *The following identities for norms of spin-preserving states hold. In each case the indicated networks are part of larger networks that are otherwise identical.*

(i) $||\sigma||=||\sigma'||$

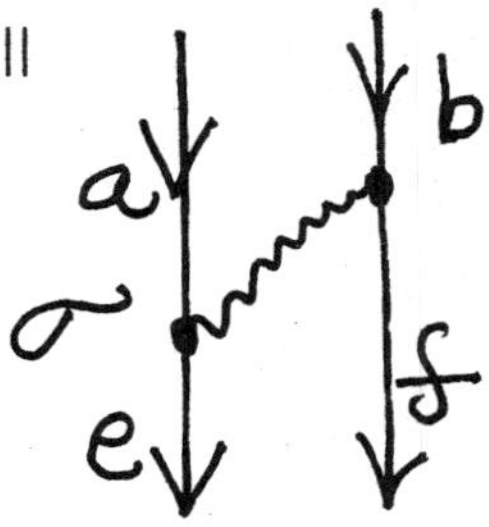

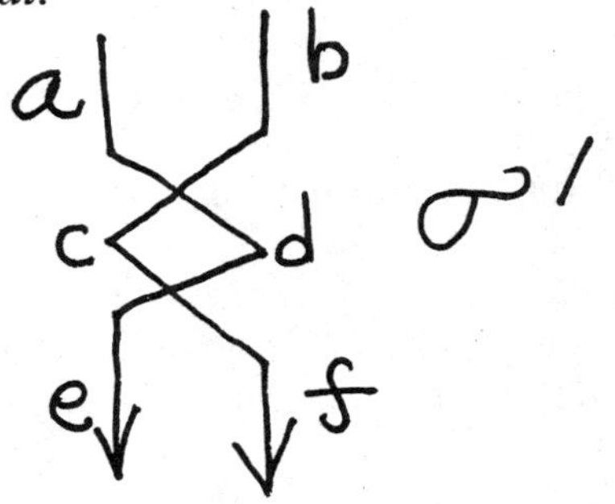

(ii) $||\sigma'|| - ||\sigma|| = [(c-b) + (f-d)]/2$

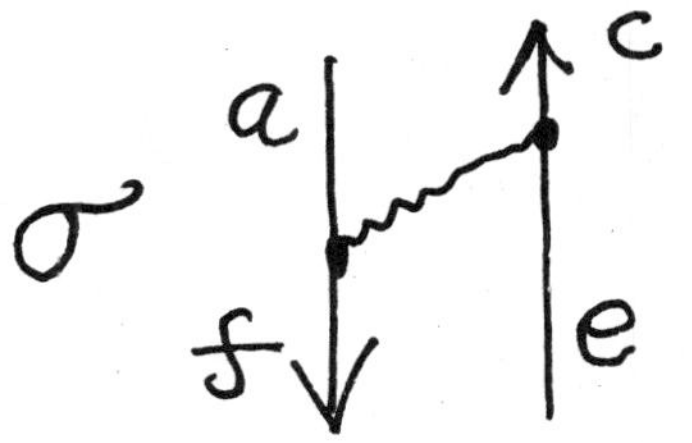

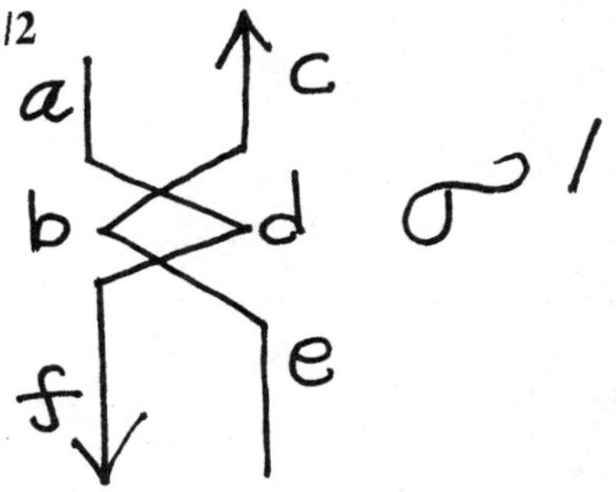

Proof. See [39] pp. 243. //

Theorem 9.6. *If the matrix*

$$S = S^{ab}_{cd}(\theta) = R^{ab}_{cd}\lambda^{(d-a)\theta/2\pi}$$

satisfies the Quantum Yang Baxter Equation (including the rapidity), then the PL state model , <K>, defined in this section, is a regular isotopy invariant of K whenever R satisfies the equations:

$$\sum_{ij} R^{ia}_{jc}(R^{-1})^{jb}_{id}\lambda^{((c-i)/2)+((d-j)/2)} = \delta^{a}_{d}\delta^{b}_{c}.$$

Proof. Apply Lemma 9.5 to conclude that the model is invariant under type IIA moves without the hypothesis of cross-channel inversion, and that cross channel inversion as phrased above insures invariance under the IIB move. Now use the results of section 2. This completes the proof. //

10. Oriented Amplitudes

Here we construct a variation on previous methods of presenting oriented models. This variation makes it possible to prove that the models of the previous section are special cases of a model for quantum amplitudes. In this section we first sketch the construction of oriented amplitudes, and then make the comparison with the models of Section 9.

In Section 6 we sketched briefly unoriented models and their relation to the bracket state sum and to quantum groups. In that model, the link diagram is arranged with respect to a vertical direction in the plane so that the critical points of the projection to this direction are maxima, minima and the crossings of the diagram. In this way the diagram is seen to decompose into cups (minima), caps (maxima) and crossings. A cup or a cap terminates in two nodes on the diagram and a crossing has four nodes. By placing these nodes on the diagram, we decompose the diagram into a composition of cups, caps and crossings. By associating abstract tensor symbols to these parts we can encode the diagram as a symbolic tensor. A cup is symbolized by a tensor with two upper indices such as M^{ab}, and a cap is symbolized by a tensor with two lower indices such as M_{ab}. The crossings are symbolized by $R^{ab}{}_{cd}$ and $(R^{-1})^{ab}{}_{cd}$. If the abstract M's and R's are replaced by actual matrices with values in a commutative ring (such as the complex numbers) and the subscripts and superscripts are interpreted as indices for these matrices (with individual indices running over a specific index set S), then the abstract tensor expression for a link diagram can be interpreted as a sum of products of elements of the ring. Here we use the Einstein summation convention and sum over repeated instances of a given index. The scenario of section 6 describes this procedure for unoriented links.

The generalization of section 6 to oriented link diagrams naturally involves the introduction of right and left oriented caps and cups. These are drawn as shown below.

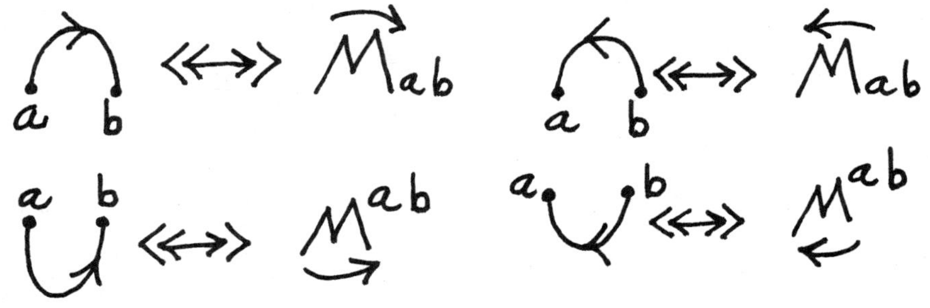

A right cup cancels with a right cap to produce an upward pointing identity line.

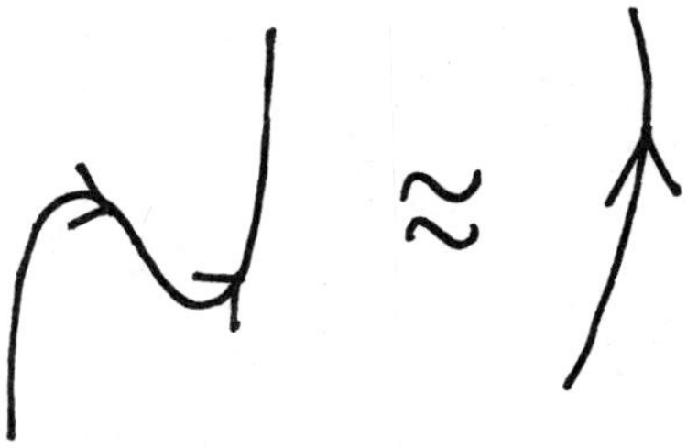

A left cup cancels with a left cap to produce a downward pointing identity line.

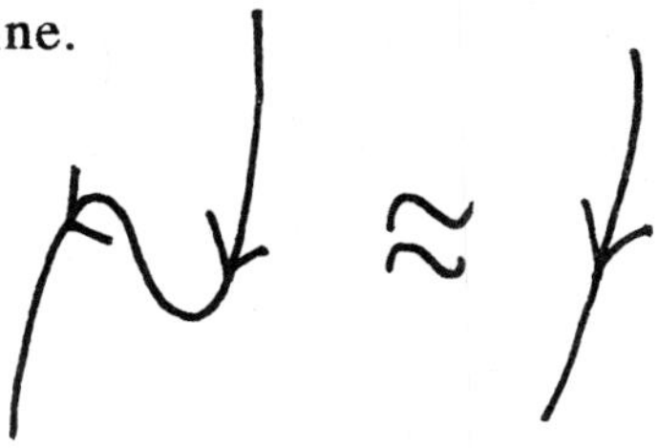

In general we need two pairs of inverse matrices for cups and caps, and there will be interrelationships with the R-matrices corresponding to the crossings in the link diagram. Just as we considered the simplifications that occur in the unoriented model by taking the cup and cap matrices to be identical (Section 6) , lets assume here that *right caps are identical with left cups* and that consequently *left caps are identical with right cups.* In fact, let us assume that the right cap and left cup are given by the matrix

$$M_{ab} = \lambda^{a/2} \delta_{ab}$$

where λ is a constant to be determined by the situation, and δ_{ab} denotes the Kronecker delta. Then the left cap and right cup are given by the inverse of M:

$$(M^{-1})_{ab} = \lambda^{-a/2} \delta_{ab}.$$

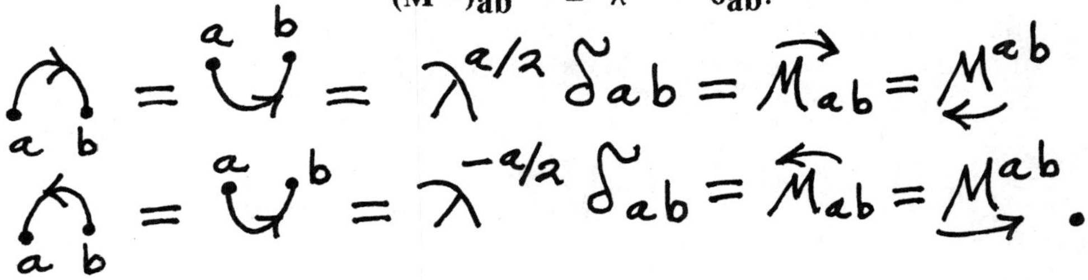

We assume that along with M we are given a solution R to the Yang-Baxter equation, and that in an oriented diagram the specific choice of $R^{ab}{}_{cd}$ is governed by the local orientation of the crossing in the diagram. Thus a and b are the labels on the lines going into the crossing and c and d are the labels on the lines emanating from the crossing.

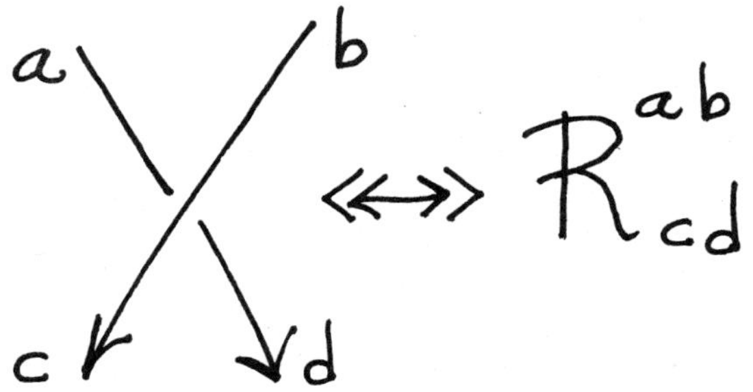

We note that with respect to the vertical direction for the amplitude, the crossings can assume the aspects: both lines pointing upward, both lines pointing downward, one line up and one line down (two cases).

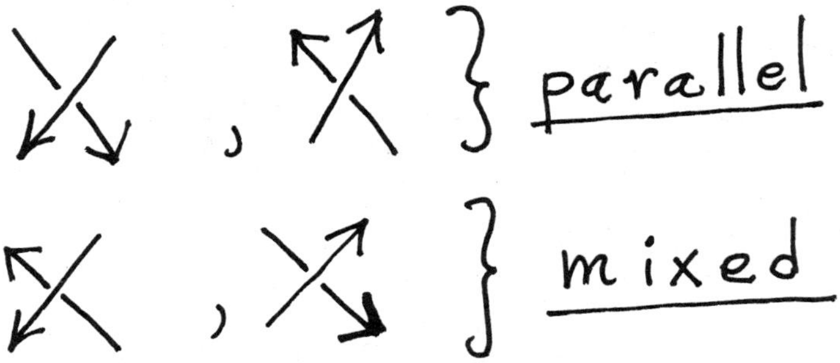

Call the cases of one line up and one line down the *mixed* cases and the upward and downward cases the *parallel* cases. A given mixed crossing can be converted, in two ways, into a combination of a parallel crossing of the same sign plus a cup and a cap. See the figure below.

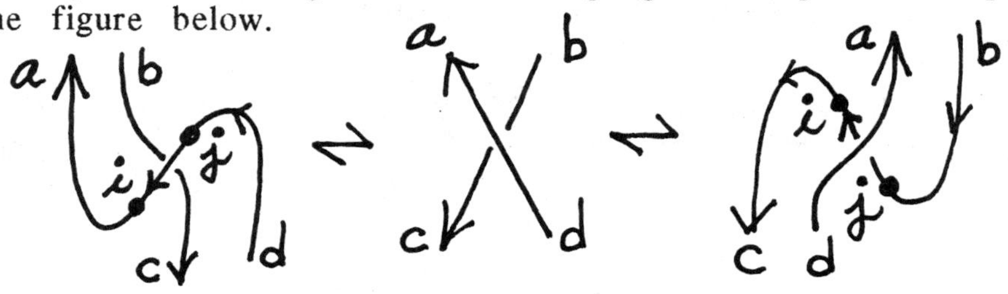

This leads to an equation that must be satisfied by the R matrix in relation to powers of λ (again we use the Einstein summation convention):

$$\lambda^{a/2}\delta_{ai}R^{bj}{}_{ic}\lambda^{-d/2}\delta_{jd} \quad = \quad \lambda^{-c/2}\delta_{ic}R^{jd}{}_{ai}\lambda^{b/2}\delta_{jb}.$$

This simplifies to the equation

$$\lambda^{a/2}R^{bd}{}_{ac}\lambda^{-d/2} \quad = \quad \lambda^{-c/2}R^{bd}{}_{ac}\lambda^{b/2}$$

from which we see that $R^{bd}{}_{ac}$ is necessarily equal to zero unless **b+d=a+c.** We say that the R matrix is *spin preserving* when it satisfies this condition. Assuming that the R matrix is spin preserving, we know from Section 2 that the model will be invariant under all orientations of the second and third Reidemeister moves just so long as it is invariant under the anti-parallel version of the second Reidemeister move:

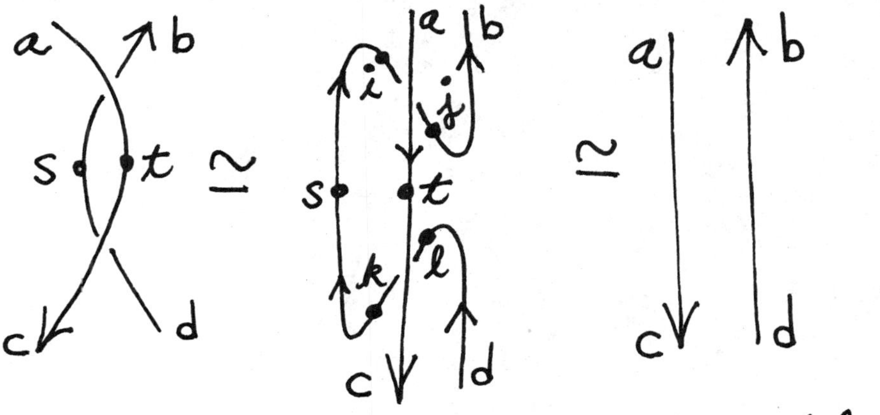

$$\overrightarrow{M}_{si}\underleftarrow{M}^{sk}\overleftarrow{M}_{\ell d}\underrightarrow{M}^{jb}R^{ia}_{tj}\bar{R}^{t\ell}_{kc}$$
$$= \delta^a_c\,\delta^b_d.$$

This antiparallel version of the second Reidemeister move places the following fundamental demand on the relation between λ and R:

$$\sum_{st} \lambda^{(s-b)/2}\lambda^{(t-c)/2} R^{sa}{}_{tb} (R^{-1})^{td}{}_{sc} = \delta^{a}{}_{c}\delta^{b}{}_{d}.$$

Call this the *R-λ equation.* The reader familiar with [30] or with the piecewise linear version as described in the previous section will recognise this equation as the fundamental requirement for regular homotopy invariance in these models. The piecewise linear model of Section 9 is not defined as an amplitude. In the piecewise linear model there is no preferred direction in the plane; rather, every crossing has an angular contribution along with its R matrix, and every angular turn in the diagram has a vertex weight corresponding to its angle and the value of the labels on its edges.

In analyzing the piecewise linear model, one shows that when the R matrix is spin-preserving, and the R-λ equation is satisfied, then the model is invariant under piecewise linear regular isotopies of the diagram. This means that one can arrange the diagram so that it is lined up with respect to a given vertical direction, and so that all the crossings are either parallel or antiparallel, and so that the angle of a parallel crossing is brought to the limiting value of *zero degrees* while the value of the angle of an antiparallel crossing is brought to a limiting value of *180 degrees* , while all other angles are right angles. These collect into pairs and produce cups and caps as pairs of right angle turns.

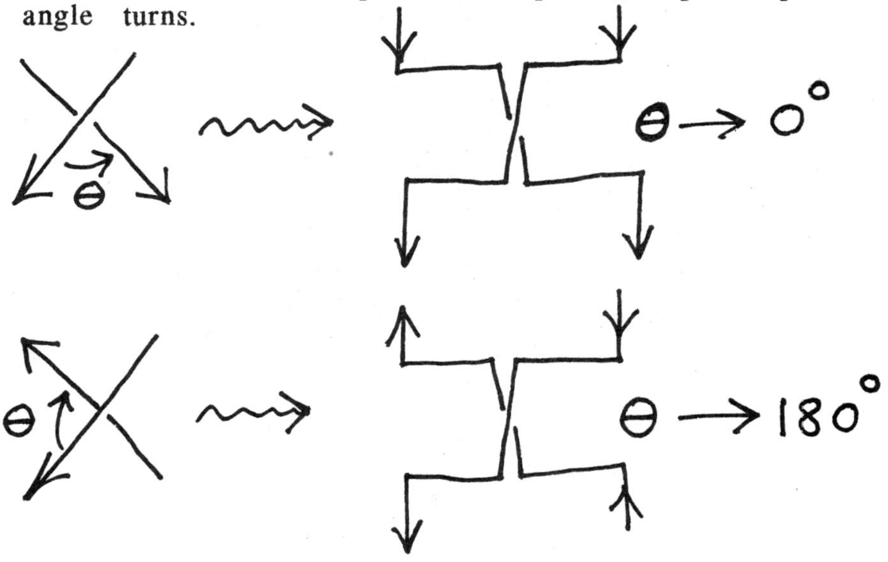

By this method the piecewise linear evaluation turns into the oriented amplitude that we have just discussed! This is a case where properties of the actual geometry of the diagrams effect the translation. The amplitude model for these invariants makes the logic of spin preservation and the R-λ equation transparent.

References

1. Y.Akutsu and W.Wadati. Knot invariants and critical statistical systems. J.Phys.Soc.Japan 56 (1987) 839-842.

2. Y.Akutsu and W.Wadati. Exactly solvable models and new link polynomials. I. N-state vertex models. J.Phys.Soc.Japan 56 (1987) 3039-3051

3. Y.Akutsu,T.Deguchi and W.Wadati. Exactly solvable models and new link polynomials II. Link polynomials for closed 3-braids, J.Phys.Soc.Japan 56 (1987) 3464-3479.

4. Y.Akutsu and M.Wadati. Knots,links,braids and exactly solvable models in statistical mechanics. Comm.Math.Phys. 117 (1988) 243-259.

5. M. F. Atiyah, *The Geometry and Physics of Knots*. Cambridge University Press, Cambridge (1990).

6. J.W.Alexander.Topological invariants of knots and links.Trans.Amer.Math.Soc. 20 (1923) 275-306.

7. R.J.Baxter. *Exactly Solved Models in Statistical Mechanics*, Academic Press (1982).

8. J.S.Birman,H.Wenzl. Braids,link polynomials and a new algebra. Trans. Amer. Math. Soc. 313 (1989) 249-273.

9. R.D.Brandt,W.B.R.Lickorish,K.C.Millett. A polynomial invariant for unoriented knots and links. Invent.Math. 84 (1986) 563-573.

10. G.Burde and H.Zieschang. *Knots* . deGruyter(1986).

11. P.N.Burgoyne.Remarks on the combinatorial approach to the Ising problem. J.Math.Phys. Vol.4, No.10,(1963).

12. J.H.Conway.An enumeration of knots and links and some of their algebraic properties. *Computational Problems in Abstract Algebra* . Pergamon Press,New York(1970),329-358.

12(a).P.Cotta-Ramusino, E.Guadagnini, M. Martellini and M.Mintchev. Quantum field theory and link invariants. (preprint 1989).

13. R.H.Crowell,R.H.Fox. *Introduction to Knot Theory* .Blaisdell Pub. Co. (1963).

14. V.G.Drinfeld. Quantum Groups, *Proc.Intl.Congress Math.*, Berkeley, Calif.USA(1986)789-820.

15. V.G.Drinfeld. Hopf algebras and the quantum Yang-Baxter equation. Soviet Math. Dokl. Vol. 32 (1985) No.1.

16. L.D.Faddeev,N.Yu.Reshetikhin,L.A.Takhtajan. Quantization of Lie groups and Lie algebras. LOMI Preprint E-14-87, Steklov Mathematical Institute, Leningrad, USSR.

17. P.Freyd,D.Yetter,J.Hoste,W.B.R.Lickorish,K.C.Millett,A.Ocneanu [HOMFLY].A new polynomial invariant of knots and links. Bull.Amer.Math.Soc. 12 (1985) 239-246.

18. J.Frohlich and C.King. The Chern-Simons theory and knot polynomials. Commun. Math. Phys. Vol. 126 (1989) 167-199.

19. J.Frohlich and C. King. Two dimensional conformal field theory and three-dimensional topology. Int. J. Mod. Phys. A, Vol.4 (1989) 5321-5399.

20. P.deLa Harpe, M.Kervaire, and C.Weber. On the Jones polynomial. L'Enseign.Math. 32 (1986) 271-335.

20.1. P. de La Harpe. Spin models for link polynomials, strongly regular graphs and Jaeger's Higman-Sims model. Pacific J. Math. Vol. 162, No. 1, (1994) 57-96.

21. R.Hartley. Conway potential functions for links, Comment.Math.Helv. 58 (1983) 365-378.

21.1 M. A. Hennings, Hopf algebras and regular isotopy invariants for link diagrams, Math. Proc. Camb. Phil. Soc. , Vol. 109 (1991) 59-77.

22. C.F.Ho.A new polynomial invariant for knots and links - preliminary report.AMS Abstracts, Vol.6 No.4,Issue 39(1985) 300.

23. J.Hoste. A polynomial invariant of knots and links. Pacific J. Math. 124 (1986) 295-320.

24. F.Jaeger. Strongly regular graphs and spin models for the Kauffman polynomial. Geom. Dedicat. Vol. 44, (1992) 23-52.

25. M.Jimbo. A q-difference analogue of U(q) and the Yang-Baxter equation. Lect. in Math. Physics 10 (1985) 63-69.

26.M.Jimbo. Quantum R-matrix for the generalized Toda system. Comm.Math.Phys. 102 (1986) 537-547

27. V.F.R.Jones. A new knot polynomial and von Neumann algebras. Notices of AMS 33 (1986) 219-225.

28. V.F.R.Jones.A polynomial invariant for links via von Neumann algebras.
Bull.Amer.Math.Soc. 129 (1985) 103-112.

29. V.F.R.Jones.Hecke algebra representations of braid groups and link
polynomials.Ann. of Math. 126 (1987) 335-388.

30. V.F.R.Jones.On knot invariants related to some statistical mechanics models. Pacific J. Math. , Vol. 137, No.2 (1989) 311-334.

31. L.H.Kauffman.The Conway polynomial.Topology 20 (1980) 101-108.

32. L.H.Kauffman. *Formal Knot Theory* . Princeton University Press Mathematical Notes #30 (1983).

33. L.H.Kauffman. *On Knots.* Annals of Mathematics Study 115, Princeton University Press (1987).

34. L.H.Kauffman. State Models and the Jones Polynomial. Topology 26 (1987) 395-407.

35. L.H.Kauffman.Invariants of graphs in three-space.
Trans.Amer.Math.Soc.Vol.311 2 (1989) 697-710.

36. L.H.Kauffman.New invariants in the theory of knots. (lectures given in Rome,June 1986.) - Asterisque, Vol.163-164 (1988)137-219.

37. L.H.Kauffman. New invariants in the theory of knots. Amer. Math. Monthly ,Vol.95,No.3,March (1988) 195-242.

38. L.H.Kauffman.An invariant of regular isotopy. Trans. Amer. Math. Soc., Vol. 318, No. 2, (1990) 417-471.

39. L.H.Kauffman. *Knots and Physics.* World Scientific Pub. (1991 and 1993).

40. L.H.Kauffman.Statistical mechanics and the Jones polynomial. In *Proceedings of the 1986 Santa Cruz conference on Artin's Braid Group.* AMS Contemp. Math. Series Vol. 78 (1989) 263-297..

41.L.H.Kauffman. A Tutte polynomial for signed graphs. Discrete Applied Mathematics. Vol. 25 (1989),pp. 105-127.

42.L.H.Kauffman. State models for knot polynomials - an introduction. (in the proceedings of the meeting of the Brasilian Mathematical Society - July 1987).

43.L.H.Kauffman. State models for link polynomials. L'Enseignment Math. Vol. 36 (1990) 1-37.

44 . L.H. Kauffman. Knots, abstract tensors and the Yang-Baxter Equation. In *Knots, Topology and Quantum field Theories.* Edited by L. Lussana, World Scientific Pub. (1989) 179-334.

44.1. L.H.Kauffman. Knot polynomials and Yang-Baxter models. *IXth International Congress on Mathematical Physics - 17-27 July-1988-Swansea, Wales,* Ed. Simon, Truman, Davies, Adam Hilger Pub. (1989) 438-441.

44.2. L.H.Kauffman. Polynomial Invariants in Knot Theory. *Braid Group, Knot Theory and Statistical Mechanics.* Ed. C.N.Yang and M.L.Ge. World Sci. Pub. Advanced Series in Mathematical Physics, Vol. 9. (1989) 27-58.

44.3. L.H.Kauffman. Spin networks and knot polynomials. Intl. J. Mod. Phys. A. Vol. 5 (1990) 93-115.

44.4. L.H. Kauffman. From knots to quantum groups and back. In *Quantum Groups.* World Sci. Pub. (1991) 1-32.

44.5. L.H. Kauffman and P. Vogel. Link polynomials and a graphical calculus. JKTR Vol.1(1992) 98-114.

44.6. L.H. Kauffman and R. Baadhio. *Quantum Topology.* World Sci. Pub. (1993).

44.7. L.H. Kauffman and S. Lins. *Temperley Lieb Recoupling Theory and Invariants of 3-Manifolds.* Annals Study No. 134. Princeton University Press (1994).

44.8. L.H. Kauffman. Gauss codes, quantum groups and ribbon Hopf algebras. Reviews in Math. Phys., Vol.5, No. 4, (1993) 735-773.

44.9. L. H. Kauffman and D. Radford. Invariants of links and 3-manifolds derived from finite dimensional Hopf algebras. Journal of Knot Theory and Its Ramifications, Vol. 4, No.1 (1995) 131-162.

45. T.P.Kirkman.The enumeration,description and construction of knots with fewer than 10 crossings. Trans. Royal Soc. Edin. 32 (1865) 281-309.

46. T.Kohno.Monodromy representations of braid groups and Yang-Baxter equations. Ann.Inst.Fourier,Grenoble 37 ,4 (1987) 139-160.

47. P.P.Kulish,N.Yu.Reshetikhin and E.K.Sklyanin. Yang-Baxter equation and representation theory:I. Letters in Math. Physics 5 (1981) 393-403.

48. A.Kuniba,Y.Akutsu,and M.Wadati. Virasoro algebra, von Neumann algebra and critical eight-vertex SOS models. J. Phys. Soc. Japan 55 , No.10 (1986) 3285-3288.

49. R.Lawrence. A universal link invariant using quantum groups. In *The Interface of Mathematics and Particle Physics.* edited by Quillen, Segal and Tsou. Oxford University Press (1990) 151-156.

50. W.B.R.Lickorish and K.C.Millett. A polynomial invariant for oriented links. Topology 26 (1987) 107-141.

51. A.S.Lipson. An evaluation of a link polynomial. Math.Proc.Camb.Phil.Soc. 100 (1986) 361-364.

52. A.S.Lipson.Some more states models for link invariants. Pacific J. Math. Vol. 152, No.2 (1992) 337-346.

53. C.N.Little.Non-alternate +- knots. Trans. Royal Soc. Edin. 35 (1889) 663-664.

54. S.Majid. Quasitriangular Hopf algebras and Yang-Baxter equations. Int. J. Mod. Phys. A, Vol. 5, No. 1 (1990) 1-91.

55. Yu.I.Manin. Quantum groups and non-commutative geometry. (Lecture Notes, Montreal, Canada , July 1988).

56. Yu.I.Manin. Some remarks on Kozul algebras and Quantum groups. Ann.Inst.Fourier.,Grenoble 37 ,4 (1987) 191-205.

57. R.Penrose.Applications of negative dimensional tensors. In *Combinatorial Mathematics and its Applications.* Edited by D.J.A.Welsh.Academic Press (1971).

58. K.Reidemeister. *Knotentheorie.* Chelsea Publishing Co.,New York (1948) Copyright 1932. Julius Springer,Berlin.

59. N.Y.Reshetikhin. Quantized universal enveloping algebras, the Yang-Baxter equation and invariants of links, I and II. LOMI reprints E-4-87 and E-17-87, Steklov Institute, Leningrad, USSR.

60. D.Rolfsen. *Knots and Links.* Publish or Perish Press (1976).

61. M.Rosso. Groupes quantiques et modeles a vertex de V. Jones en theorie des noeuds. C.R. Acad. Sci. Paris, t. 307, Serie I, (1988) 207-210.

62. C.Schultz. Solvable q-state models in lattice statistics and quantum field theory. Phys.Rev.Lett. 46 No.10,March 1981.

63. P.G.Tait.On Knots I,II,III.Scientific Papers Vol.I,Cambridge University Press.,London,(1898) 273-347.

64. H.N.V. Temperley and E.H. Lieb. Relations between the percolation and coloring problem and other graph theoretical problems associated with regular planar lattice: some exact results for the percolation problem. Proc. R. Soc. London A Vol. 322 (1971) 251-280.

65. M.B.Thistlethwaite. Knot tabulations and related topics. *Aspects of Topology.* Ed. I.M.James and E.H.Kronheimer, Cambridge University Press (1985) 1-76.

66. B.Trace. On the Reidemeister moves of a classical knot. Proc.Amer.Math.Soc. Vol.89, No.4 (1983).

67. V.G.Turaev. The Yang-Baxter equations and invariants of links. LOMI preprint E-3-87, Steklov Institute, Leningrad, USSR. Inventiones Math. 92 (1988) 527-553.

68. H.Whitney.On regular closed curves in the plane. Comp.Math. 4 (1937) 276-284.

69. E.Witten. Physics and geometry. Proc. Intl. Congress Math., Berkeley, Calif.USA (1986)267-303.

70. E.Witten. Quantum field theory and the Jones polynomial. Commun.Math.Phys. 121 (1989) 351-399.

71. F.Y. Wu. Knot theory and statistical mechanics. Reviews of Modern Physics. Vol. 64, No. 4, (1992).

72. F.Y. Wu, P.Pant and C. King. A new link invariant from the Chiral Potts model. (preprint 1994).

73. A.B.Zamolodchikov. Factorized S matrices and lattice statistical systems. Soviet Sci. Reviews. PartA (1979-1980).

72. A.B.Zamolodchikov. Tetrahedron equations and the relativistic S-matrix of straight strings in 2+1 dimensions. Comm.Math.Phys. 79 (1981) 489-505.

Department of Mathematics, Statistics and Computer Science
The University of Illinois at Chicago
851 S. Morgan St.
Chicago, Illinois 60607-7045
USA

Proceedings of Symposia in Applied Mathematics
Volume **51**, 1996

An Introduction to Topological Field Theory

R.J. Lawrence

Abstract. A topological quantum field theory (TQFT) is an, almost, metric independent quantum field theory that gives rise to topological invariants of the background manifold. The best known example of a 3-dimensional TQFT is Chern-Simons-Witten theory, in which the expectation value of an observable, obtained as the product of the Wilson loops associated with a link, is the generalised Jones invariant of the link. Unfortunately the form of the invariants obtained by this procedure is that of an integral over an infinite dimensional space on which, for a mathematician, a measure has not yet been rigorously defined. Various ways of avoiding this difficulty have been developed. These fall into two main categories, namely, formal manipulations of Witten's path integral into a form which can then be rigorously defined, and axiomatic encapsulations of the properties of TQFTs. In these notes we will be concerned with the second path, demonstrating how complex categorical and algebraic structures appear, from apparently simple geometry. These structures are related to the quantum group structures which arise in other approaches.

1: Introduction

In these notes an elementary introduction to some of the basic algebraic structures arising in topological field theory (TFT) will be given. The case of two dimensional topological field theories will be covered in detail. Although the classification of manifolds in two dimensions is very simple, the analysis of possible theories is yet interesting because, when correctly interpreted, it contains in a simple form all the main features which arise in higher dimensional theories. In §2, the basic axioms for TFTs will be formulated, stimulated by structures which arise in Quantum Field Theory and Statistical Mechanics. The associated geometrically motivated structure of a domain category is used in §2.4 to express a TFT as a functor from a domain category of manifolds to a domain category of vector spaces. This leads to notions of TFTs on manifolds endowed with various additional structures, while in §2.5 a technique for generating a TFT over a class of such manifolds, from one on manifolds with even more additional structure, is given. In §3, the process of §2 is seen in action, examples being generated of TFTs, both over the collection of topological manifolds and the collection of triangulated manifolds.

The extension of these ideas to higher dimensions are briefly discussed in §4 and §5 where general notions of higher category structure and extended topological

---

1991 *Mathematics Subject Classification.* Primary 81T40; Secondary 18F99, 57Q45.

The author is a Raymond and Beverly Sackler Fellow at IHES, France.

This paper is in final form and has not been submitted for publication elsewhere.

field theories (ETFTs) are introduced. Some relationships with constructions of invariants of links and 3-manifolds are also mentioned.

Since the field is a huge and rapidly increasing one, these notes necessarily only touch on the beginnings of the field and leave many loose ends; however, it is hoped that they will give the reader a feel for the origins of some of the structures naturally arising in TFT.

2: Formulation of axioms

2.1 Motivation

In classical mechanics, there are two ways of describing the possible evolutions of a system. The first is to specify the equations of motion, which will determine the state of a system at any time (as a point in phase space) once it is given at some initial time; this is known as the *Hamiltonian approach*. The second description proceeds on the assumption that the worldline followed is one which minimises a certain functional, known as the *action*. This minimisation takes place over all possible paths in the configuration space, beginning and ending at given points, and is known as the *Lagrangian approach*. In quantum theory, the phase space is replaced by a Hilbert space of possible states, and dynamical variables are replaced by observables, which are operators on the Hilbert space and have expectation values. In a quantum field theory, the states of the system studied are specified by fields on the background manifold. The Hamiltonian approach leads to the consideration of operators on Hilbert space which describe the evolution of the state of the system. In the Lagrangian approach, the basic object which arises is the *partition function* of the theory which can be expressed as a Feynman integral.

As for any quantum theory, the ouput from a quantum field theory is a collection of expectation values and correlation functions of observables. A *topological* field theory is a theory in which the output is unchanged under a variation of the metric on the background manifold, so that expectation values of observables must give rise to topological invariants of the manifold.

Jones-Witten theory

The first interesting topological field theory was introduced by Witten in 1988, see [W].[1] The partition function of the theory supplies invariants of 3-dimensional manifolds in the form of a Feynman integral,

$$Z(M) = \int_{\mathcal{A}} e^{ikCS(A)}\mathcal{D}A\,, \tag{2.1.1}$$

where $CS(A) = \frac{1}{4\pi}\int_M \langle A, dA + 1/3[A,A]\rangle$ is the Chern-Simons action.[2] The data for this theory consists of an integer k, called the level, and a Lie group, G. Here the

[1] Abelian Chern-Simons theory had already been investigated by A.S. Schwarz [Sch1], a decade earlier while he already proposed [Sch2] in 1987 that the Jones polynomial could arise as a topological invariant from Chern-Simons quantum field theory.

[2] Here, A is a $\mathfrak{g}$-valued 1-form on M, $[\cdot,\cdot]$ is the Lie bracket and $\langle\cdot,\cdot\rangle$ is an invariant bilinear form on $\mathfrak{g}$. Thus dA and $[A,A]$ are $\mathfrak{g}$-valued 2-forms and the integrand of $CS(A)$ is therefore a scalar-valued 3-form on M which it thus makes sense to integrate over M, giving the scalar action functional $CS(A)$.

integral for $Z(M)$ takes place over the quotient by the action of a gauge group, of the space of all G–connections on a principal G–bundle over M, that is, of $\mathfrak{g}$–valued 1–forms, A, on M. For large k, the major contribution to $Z(M)$ comes from those connections A close to those at which $CS(A)$ is stationary, that is, at which the first variation of $CS(A)$ vanishes; this is the so-called stationary phase approximation. However, $\frac{\delta CS(A)}{\delta A}$ is, up to a scalar factor, just the curvature of A. Therefore, the main contribution to $Z(M)$ for large k comes from flat connections; these are the 'classical' states for this system.

From the same field theory, Witten also generated invariants, $Z(M, L)$, of pairs (M, L), where L is a link embedded in M, as the expectation value of a suitable observable known as a *Wilson loop*. Additional data of a choice of representation, ρ_i, of G for each component of L is needed, and then one defines,

$$Z(M, L) = \int_{\mathcal{A}} e^{ikCS(A)} \prod_{L_i} \mathrm{tr}_{\rho_i} \left(P \exp \oint_{L_i} A ds \right) \mathcal{D}A \,, \tag{2.1.2}$$

where the product is over all components L_i of L. The integrand in (2.1.2) differs from that in (2.1.1) by the insertion of a product of terms (Wilson loops[3]), one associated with each component of the link L, each of which may be geometrically expressed as the trace of the holonomy of the connection A around that component.[4]

Once the Lie group is fixed, the above invariants are complex-valued, depending upon the choice of the integer[5] simplest case of $G = SL(2)$, these Witten invariants for links in S^3, all of whose components are labelled by the 2–dimensional representation, reproduce the one-variable Jones' polynomial, an invariant taking values in polynomials in q, which was discovered in 1984 in quite a different way, see [J1]. When $G = SL(m)$ and the components are all labelled by the vector representation, the invariant obtained generate a slice of the 2–variable FLYPMOTH[6] polynomial[7],

---

3 Strictly speaking, a Wilson loop is associated only with a single loop in M; however, by an abuse of terminology, we will use the term 'Wilson loop associated with L' to mean the product of the Wilson loops associated with the components of L.

4 To understand the use of the term expectation value in this context, one must think of the integrand in (2.1.1) as a (complex-valued, unnormalised) probability measure on the space of quantum states, $\mathcal{A}$. The expectation value, as defined in probability theory, when applied to the Wilson loop associated with a link L, is now $Z(M,L)/Z(M)$.

5 The necessity for k to be integral eminates from the fact that $CS(A)$ is not invariant under the action of the gauge group, but rather may change by the addition of integral multiples of 2π. Therefore, only for integral k is the integrand of (2.1.2) well-defined on $\mathcal{A}$.

6 Named after the initials of the authors who, working in five independent groups, (almost) simultaneously discovered this generalisation of the one-variable Jones polynomial. Four of the groups sent papers to Bull. A.M.S., which were published in a combined and shortened form; consequently the invariant is commonly referred to by the letters HOMFLY. The arrangement of the letters does not reflect priority or indicate those working in pairs. Although C.F. Gauss and W. Listing had both investigated knot theory, the first serious study of the subject was undertaken by P.J. Tait, resulting from the hoped for application to physics related to the vortex theory of the atom proposed by Sir William Thompson (later Lord Kelvin) and James Clark Maxwell. My preference for the name FLYPMOTH is partially due to Tait's use of the Scots word 'flype' (whose best English equivalent is 'turning outside in' according to Tait) and partially because it then includes the contribution of [PT]. 'Flype' describes an operation on knots which is central to his famous set of conjectures, known as the Tait conjectures and not proved until the last decade, with the advent of the new 'quantum' link invariants.

7 The value obtained is, up to normalisation, $X_L(q, q^{m-1})$ where X_L is an invariant of links taking values which are polynomials in two variables. The fact that in these cases the invariant is actually a polynomial in q and is therefore well-defined even when q is not a root of unity, or equivalently k is not an integer, may seem surprising. It turns out that this holds for all Jones-Witten invariants of links in S^3, but currently all constructions of these invariants for general 3–manifolds, with or without embedded links, require q to be a root of unity. The reasons for this are deep and related to the representation theory of quantum groups; I believe that such restrictions on q will soon be shown to be unnecessary.

see [HOMFLY], [PT] and [J2]. These invariants were originally constructed combinatorially from presentations of links, either in the form of two–dimensional projections with over and under crossings, or as closures of braids. These are somewhat unsatisfactory formulations since the invariance of the result is not immediately evident. On the other hand, Witten's functional integral can be fairly easily seen, by a (non-rigorous) formal 'proof', to define a topological invariant. However, it has the disadvantage of not being defined rigorously because it is unclear what measure, $\mathcal{D}A$, may be placed on the infinite dimensional space $\mathcal{A}$. Various ways of avoiding this difficulty by formal manipulations of Witten's path integral into a form which *can* be rigorously defined, have been developed.

In these notes, however, our basic philosophy is as follows. We wish to forget about the origins of these theories in physics and attempt to make sense of the functional integral form of Z. Instead of pursuing the physical connections, we try to extract properties which could be expected to be satisfied by expressions of the form of (2.1.1), if indeed any reasonable meaning could be attached to them. These properties then become our axioms and we attempt to find solutions[8] to this system of axioms by using methods which may be completely unrelated to the physical origins of (2.1.1).

From our perspective, the output of a theory is a topological invariant, $Z(M)$, defined for objects M which, for the moment, we think of as arbitrary closed oriented manifolds, of fixed dimension. In practice, objects will often be endowed with extra information, such as a triangulation, a framing or a metric, and indeed, it is not even necessary for M to have any geometric interpretation so long as there are certain formal operations defined, such as boundary and union.

Feynman type sums

To motivate the initial set of axioms for TFT, we will start by investigating expressions which look formally similar to (2.1.1). Although we have it in mind that $Z(M)$ will depend only on the structures inherent in M, it may be easier to define a scalar from M when it is endowed with extra structure. One way of obtaining a quantity independent of this structure, is to sum the values obtained over all such structures. Thus we suppose that $Z(M)$ can be expressed in the form,

$$Z(M) = \int_{\mathcal{A}(M)} W_M(A)\mathcal{D}A\,. \qquad (2.1.3)$$

Here $\mathcal{A}(M)$ represents the possible set of extra data on M; $W_M(A)$ is a scalar (the *weight*) computed from M and $A \in \mathcal{A}(M)$; while $\mathcal{D}A$ represents a measure on $\mathcal{A}(M)$. If $\mathcal{A}(M)$ happens to be a discrete set then the integral becomes a summation and the measure turns into a weighting.

Suppose further that the allowed additional data that is placed on M has the following properties.

(i) (*Restriction*) Data may be restricted to submanifolds in a well-defined way, so that there are maps $r_{N,M}\colon \mathcal{A}(M) \longrightarrow \mathcal{A}(N)$ for $N \subset M$ with the property that $r_{P,N} \circ r_{N,M} = r_{P,M}$.

[8] For the particular case of invariants of links in 3-manifolds given by (2.1.2), this was carried out in [RT] starting from a modular-Hopf algebra. These invariants have consequently come to be known as Witten-Reshetikhin-Turaev invariants.

(ii) (*Pasting*) Data on two submanifolds M_1 and M_2 may be pasted together so long as it matches on the intersection. Thus if M is split into two parts (see Figure 1), M_1 and M_2, by a codimension 1 submanifold, Σ, we are supposing that for any $A_i \in \mathcal{A}(M_i)$ $(i = 1, 2)$ with $r_{\Sigma,M_1}(A_1) = r_{\Sigma,M_2}(A_2)$, there exists a unique $A \in \mathcal{A}(M)$ such that $A_i = r_{M_i,M}(A)$ for $i = 1, 2$.

(iii) (*Multiplicativity*) The weight can be defined on manifolds with boundary and is multiplicative under pasting. That is, in the situation of (ii), $W_M(A) = W_{M_1}(A_1).W_{M_2}(A_2)$.

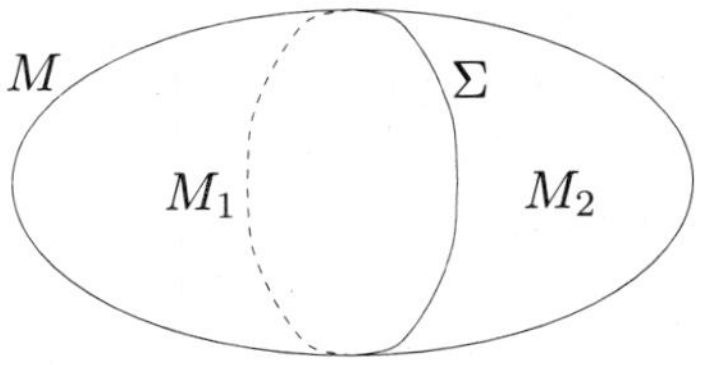

Figure 1: Splitting of a closed manifold

Given boundary data $A_0 \in \mathcal{A}(\Sigma)$, for any manifold M with boundary Σ, we let $\mathcal{A}(M, A_0) \subset \mathcal{A}(M)$, denote the set of all data on M which restricts to A_0 on Σ. The above properties of $\mathcal{A}$ and W enable $Z(M)$, where M is obtained by gluing together two pieces M_1 and M_2 along their common boundary as in (ii), to be written as

$$Z(M) = \sum_{A_0 \in \mathcal{A}(\Sigma)} Z(M_1, A_0) Z(M_2, A_0) , \tag{2.1.4}$$

where $Z(M_i, A_0)$ is the part of the sum in (2.1.3) consisting of those terms for which $A \in \mathcal{A}(M_i, A_0)$. Equation (2.1.4) may be rewritten as,

$$Z(M) = \langle Z(M_1) \mid Z(M_2) \rangle , \tag{2.1.5}$$

a contraction of two vectors $\langle Z(M_1) \mid$ and $\mid Z(M_2) \rangle$ which lie in dual[9] vector spaces, whose bases are indexed by $\mathcal{A}(\Sigma)$.

Expressions of the form (2.1.3) arise in many areas. In statistical mechanics, $\mathcal{A}(M)$ will represent a set of possible states of the system and $W_M(A)$ will be the relative weight of that state, $\mathrm{e}^{-E_M(A)/kT}$ where $E_M(A)$ is the energy associated with A, T is the temperature and k is Boltzmann's constant. In field theory, $\mathcal{A}(M)$ represents a set of possible fields on M and $W_M(A) = \mathrm{e}^{iS(A)}$, where $S(A)$ represents the action.[10] The physical meaning of the conditions (i)–(iii) above is *locality* of the theory. Thus for a statistical mechanical model on a lattice, it is the requirement that states may be specified by local pieces of data (e.g. spins at vertices), while the energy of a state may be expressed as the sum of local (nearest-neighbour) interaction energies.

---

[9] The vector spaces are dual rather than identical due to the diffence in relative orientations of Σ with respect to M_1 and M_2; such questions of orientation have been ignored in the above discussion.

[10] Usually $\mathcal{A}(M)$ is actually a quotient space by the action of an appropriate gauge group. At the level of the present discussion we ignore such complications.

2.2 Axioms

The above discussion motivates the following original definition of a $(d+1)$–dimensional TFT, which is due to Atiyah and was modelled on Segal's axioms for Conformal Field Theory (see [S1] and [S2]). The theory should assign,

(a) to every closed d–dimensional manifold, Σ, a complex vector space which will be denoted by $Z(\Sigma)$;

(b) to every $(d+1)$–dimensional manifold, M, a vector $Z(M) \in Z(\partial M)$.

Here the manifolds may be endowed with additional structure and/or constraints; it is assumed that all manifolds are oriented. We generally use the symbols M, Σ and C with appropriate suffices to refer to manifolds of codimension 0,1 and 2, respectively, relative to the dimension of the theory under consideration (here $d+1$). The items $Z(M)$ and $Z(\Sigma)$ should satisfy the axioms below.

A1. (VACUUM) $Z(\emptyset) = \mathbf{C}$.

A2. (DUALITY) $Z(\Sigma^*) = Z(\Sigma)^*$.

A3. (MULTIPLICATIVITY) $Z(\Sigma_1 \amalg \Sigma_2) = Z(\Sigma_1) \otimes Z(\Sigma_2)$.

A3′. (GLUING)

(a) $Z(M_1 \amalg M_2)$ and $Z(M_1) \otimes Z(M_2)$ are equal as elements of $Z(\partial M_1) \otimes Z(\partial M_2)$.

(b) If M is a manifold with $\partial M = \Sigma_1 \amalg \Sigma \amalg \Sigma^*$ while $\cup_\Sigma M$ is the manifold obtained from M by identifying the boundary components Σ and Σ^*, then

$$Z(\cup_\Sigma M) = \circ_{Z(\Sigma)} Z(M)\,,$$

where $\circ_{Z(\Sigma)} \colon Z(\Sigma_1) \otimes Z(\Sigma) \otimes Z(\Sigma)^* \longrightarrow Z(\Sigma_1)$ is the natural contraction map on the second and third factors.

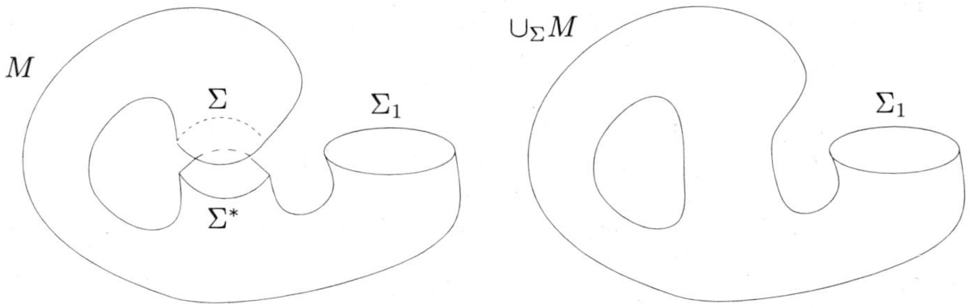

Figure 2: Gluing along a boundary component

Remark 2.2.1

(i) Topological field theories may be defined over any ring, replacing $\mathbf{C}$ in axiom A1.

(ii) In axiom A2, the symbol '$*$' is used in two different ways on the two sides of the equation. On the left hand side it refers to the operation of reversing the orientation, while on the right hand side it is that of taking the dual on vector spaces.

(iii) As a consequence of axioms A2 and A3, if M, Σ_1 and Σ_2 satisfy $\partial M = \Sigma_1^* \amalg \Sigma_2$ then the theory supplies

$$Z(M) \in Z(\partial M) = Z(\Sigma_1^* \amalg \Sigma_2) = Z(\Sigma_1)^* \otimes Z(\Sigma_2)\,,$$

which may be alternatively viewed as a map $Z(\Sigma_1) \longrightarrow Z(\Sigma_2)$, so that the theory associates linear maps to cobordisms.

(iv) With the notation of the previous remark, one may view M alternatively as a cobordism from Σ_2^* to Σ_1^*. The associated map $Z(\Sigma_2)^* \longrightarrow Z(\Sigma_1)^*$ is now the adjoint of the map $Z(M)$ of the previous remark.

(v) Suppose that M_1 and M_2 are cobordisms from Σ_1 to Σ and from Σ to Σ_2, respectively. Using axioms A3′(a) and (b) it follows that

$$\big(Z(M) = Z(M_2) \circ Z(M_1)\big)\colon Z(\Sigma_1) \longrightarrow Z(\Sigma_2)\,,$$

where M is obtained by gluing M_1 and M_2 along Σ. For this reason, axiom A3′ is often referred to as the *associativity* axiom. Equation (2.1.5) follows from the special case when $\Sigma_1 = \Sigma_2 = \emptyset$.

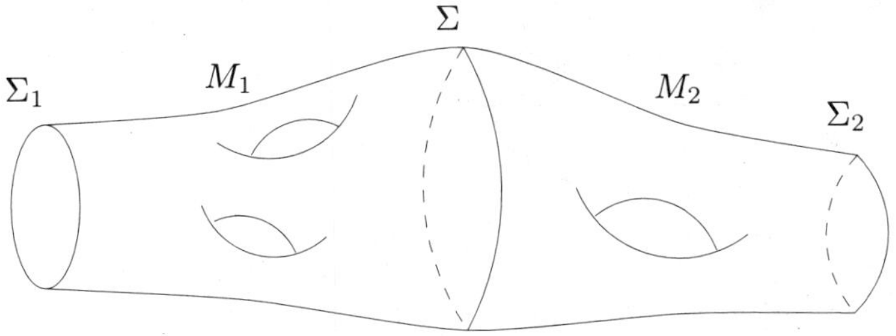

Figure 3: Gluing cobordisms

It is necessary to introduce a further subtlety since it is not really possible to decide when two manifolds are actually *equal*, the most that exists is an isomorphism between them. The same is true for vector spaces and so we arrive at the following additional axioms.

A0. (NATURALITY) Any isomorphism $f\colon \Sigma \longrightarrow \Sigma'$ of surfaces of codimension one induces an isomorphism $Z(f)\colon Z(\Sigma) \longrightarrow Z(\Sigma')$ in such a way that $Z(f') \circ Z(f) = Z(f' \circ f)$ for all $f'\colon \Sigma' \longrightarrow \Sigma''$.

A0′. (NATURALITY) For any isomorphism $f\colon M \longrightarrow M'$, the restriction to ∂M gives an induced isomorphism for which $Z(M') = Z(f|_{\partial M})Z(M)$.

Remark 2.2.2

(i) In axioms A0 and A0′, the isomorphisms, f, are to be thought of as topological equivalences of manifolds, while $Z(f)$ is an isomorphism of vector spaces.

(ii) The same symbol, Z, is here being used to specify the object assigned by the theory to several different object types. Thus $Z(*)$ is a vector space, vector, or map according as $*$ is a codimension-one object, codimension-zero object, or morphism specifying the equivalence of codimension-one objects.

(iii) In Remark 2.2.1(iii), the map $Z(M)$ also depends upon a choice of isomorphism from ∂M to $\Sigma_1^* \amalg \Sigma_2$. Strictly speaking it is therefore necessary to introduce extra isomorphisms in place of equalities in axioms A1, A2, A3 and A3′; see §2.4 where such isomorphisms are naturally contained in a suitably enhanced notion of category.

In some theories, $Z(\Sigma)$ may possess, in addition, a *– structure, such as when it is a Hilbert space. In such a situation there is a natural identification of $Z(\Sigma)$ with its dual and it then makes sense to consider a further axiom.

A2′. (CONJUGACY) $Z(M^*) = Z(M)^*$ where $Z(M^*) \in Z(\partial M)^*$ is identified with an element of $Z(\Sigma)$.

Those axioms whose labels differ only by the addition of a prime are very similar to each other, the main difference being in the dimension of the base objects, namely, codimension–0 in those axioms with primed label as opposed to codimension–1 in the other axiom. The similarities will become clearer in §2.4 and are extended in §5.

2.3 Consequences

The axioms given above, while they do not determine the theory, Z, do impose quite strong conditions on the structure of $Z(M)$ and $Z(\Sigma)$. Before moving on to a reformulation of the axioms in terms of categories in the next section, we give three immediate consequences of the axioms.

A. Firstly, if it is possible to cut up a manifold M into 'elementary pieces' by closed submanifolds of codimension–1, then the axioms may be used to evaluate $Z(M)$ in terms of the vectors associated with those pieces. This allows theories to be specified completely by a small number of pieces of data, which, as we will see in the next section, may be thought of as constituting an algebraic structure, in an appropriate sense.

B. Suppose next that Σ is a codimension–1 object. By Remark 2.2.1(iii), the cylinder $\Sigma \times I$ ($I = [0,1]$) supplies a map $i_\Sigma \colon Z(\Sigma) \longrightarrow Z(\Sigma)$. The result of gluing together two copies of this cylinder along a common boundary, is isomorphic to the original cylinder and therefore, by axiom A3′,

$$i_\Sigma \circ i_\Sigma = i_\Sigma \,.$$

Thus i_Σ is an idempotent, whose image defines a subspace of $Z(\Sigma)$ which we will denote by $Z'(\Sigma)$. According to Remark 2.2.1(iv), $i_{\Sigma^*} = (i_\Sigma)^*$ and so $Z'(\Sigma^*) \cong Z'(\Sigma)^*$. By axiom A3, $i_{\Sigma_1 \amalg \Sigma_2} = i_{\Sigma_1} \otimes i_{\Sigma_2}$, from which it follows that

$Z'(\Sigma_1 \amalg \Sigma_2) = Z'(\Sigma_1) \otimes Z'(\Sigma_2)$. For any manifold, M, it follows from axiom A3′(b), by gluing a cylinder on ∂M onto M, that,

$$i_{\partial M}(Z(M)) = Z(M) \in Z(\partial M)\,.$$

We conclude that $Z(M) \in Z'(\Sigma)$ and therefore that setting $Z'(M) \equiv Z(M)$ defines a smaller theory Z' which also satisfies the axioms for a TFT. As a result, in the classification of TFTs we may assume, without loss of generality, that $Z'(\Sigma) = Z(\Sigma)$ for all Σ. Hence we assume throughout the rest of these notes that

$$i_\Sigma = \mathrm{id}_{Z(\Sigma)}\,.$$

C. Lastly, suppose that we apply axiom A3′(b) to the cylinder $\Sigma \times I$. The result of identifying the two copies of Σ in the boundary is $\Sigma \times S^1$. Thus $Z(\Sigma \times S^1) \in Z(\emptyset) = \mathbf{C}$ is obtained as the trace of i_Σ, that is,

$$Z(\Sigma \times S^1) = \dim Z(\Sigma)\,.$$

2.4 Interpretation using categories

A neat way of encapsulating the axioms for TFTs is in terms of categories. For a nice account of category theory, see [M]. So that the reader who is unfamiliar with the notion of a category can still follow these notes, we give here the definition of a category and some elementary examples which will be relevant to us.

Categories

Definition 2.4.1 *By a category, $\mathcal{C}$, is meant a set of objects, denoted* $\mathrm{Obj}\,(\mathcal{C})$, *along with, for each pair of objects* $A, B \in \mathrm{Obj}\,(\mathcal{C})$, *a set,* $\mathrm{Morph}_{\mathcal{C}}(A,B)$, *of morphisms and the following additional structures.*

(a) For each $A \in \mathrm{Obj}\,(\mathcal{C})$, *an element* $\mathrm{id}_A \in \mathrm{Morph}_{\mathcal{C}}(A,A)$, *called the identity morphism on the object A.*

(b) For each triple $A, B, C \in \mathrm{Obj}\,(\mathcal{C})$, *a map,*

$$\circ\colon \mathrm{Morph}_{\mathcal{C}}(B,C) \times \mathrm{Morph}_{\mathcal{C}}(A,B) \longrightarrow \mathrm{Morph}_{\mathcal{C}}(A,C)$$

called composition, which is such that $\mathrm{id}_B \circ f = f = f \circ \mathrm{id}_A$ *and* $f\circ(g\circ h) = (f\circ g)\circ h$, *the equalities holding for all objects and morphisms for which both sides make sense.*

We will often use the same notation for the set of objects in a category and the category itself, so that '$a \in \mathcal{C}$' means '$a \in \mathrm{Obj}\,(\mathcal{C})$'. Also, when it is clear in which category we are working, we will drop the suffix to Morph. If A and B are two objects in the category $\mathcal{C}$, it is sometimes convenient, and easier on the reader, to represent the fact that $f \in \mathrm{Morph}\,(A,B)$ by $f\colon A \longrightarrow B$.

Example 2.4.2 If G is a group, then one may construct a category with a one-element set of objects, for which the set of morphisms from this object to itself is just G, the identity and multiplication operations in G providing the requisite structures in Definition 2.4.1(a) and (b), respectively. Observe that only the identity and associativity properties of G have been used, so that this construction works equally well for any monoid. However, in the special case where G is a group, the category obtained is a *groupoid*, that is a category in which every morphism is an equivalence.[11]

Example 2.4.3 Another way of viewing a category is as a semi-group structure in which the multiplication map is only defined on a *subset* of the set of all pairs of elements. Thus, if $\mathcal{C}$ is a category, let $C^0 = \mathrm{Obj}\,(\mathcal{C})$ and C^1 denote the (disjoint) union of the sets $\{A\} \times \{B\} \times \mathrm{Morph}\,(A, B)$ over all pairs of objects A and B. Then there are defined two maps $s, t: C^1 \longrightarrow C^0$, called the source and target maps, under which the images of (A, B, f) are A and B, respectively. Elements $a, b \in C^1$ can be composed so long as $s(a) = t(b)$. The identity morphism provides a map $i: C^0 \longrightarrow C^1$ for which $s \circ i = t \circ i = \mathrm{id}_{C^0}$, while part (b) of Definition 2.4.1 are identity and associativity constraints.

Example 2.4.4 On any set, S, there are two trivial category structures whose object set is S. For the first, $\mathrm{Morph}\,(A, B)$ is a one-element set for all $A, B \in S$, while for the second $\mathrm{Morph}\,(A, B)$ is empty, unless $A = B$, when it contains just one element. We will denote these categories by $\mathrm{ALL}(S)$ and $\mathrm{TRIV}(S)$, respectively.

Example 2.4.5 The category of vector spaces, $\mathcal{V}^1$, has as objects, finite dimensional vector spaces, and as morphisms, linear transformations. Since any finite dimensional vector space may be determined up to isomorphism by its dimension, one may also consider the category of *coordinatised* vector spaces, $\mathcal{V}^1_c$, whose set of objects is $\mathbf{N} \cup \{0\}$ and for which $\mathrm{Morph}\,(m, n)$ is the set of $n \times m$ matrices with entries in $\mathbf{C}$. Composition is given by matrix multiplication while the identity morphism on n is the $n \times n$ identity matrix.[12]

Example 2.4.6 Suppose that G is any group. Then one may define a category whose class of objects consists of pairs (V, ρ), where V is a vector space and $\rho: G \longrightarrow \mathrm{End}(V)$ is a representation of G. For two objects (V, ρ) and (W, σ), the set of morphisms is defined to be the set of linear transformations $\alpha: V \longrightarrow W$ for which $\alpha\big(\rho(g)\mathbf{v}\big) = \sigma(g)\alpha(\mathbf{v})$ for all $g \in G$ and $\mathbf{v} \in V$. The resulting category is known as the *category of representations* of G. This same construction can be used for almost any algebraic structure. The smallest structure for which it can be applied is a groupoid, while additional structure on G induces additional functors on the category.

---

[11] A morphism $f \in \mathrm{Morph}_{\mathcal{C}}(A,B)$ is said to be an equivalence if it possesses an inverse, that is, if there exists a $g \in \mathrm{Morph}_{\mathcal{C}}(B,A)$, in the sense that the composition of the two morphisms in either order produces an identity morphism on an appropriate object, $f \circ g = \mathrm{id}_B$ and $g \circ f = \mathrm{id}_A$.

[12] There is a formal notion in category theory known as a skeleton of a category (see [M]) and in this sense $\mathcal{V}^1_c$ is the skeleton of $\mathcal{V}^1$. The skeleton of $\mathrm{ALL}(S)$ is the category with one object while that of $\mathrm{TRIV}(S)$ is $\mathrm{TRIV}(S)$.

Example 2.4.7 Suppose that M is a manifold. Construct a category $\mathcal{C}$, whose set of objects is the set of points in M, as follows. If $p_0, p_1 \in M$ then $\mathrm{Morph}\,(p_0, p_1)$ consists of all homotopy classes of paths in M from p_0 to p_1. That is, a morphism is an equivalence class of maps $\gamma\colon [0,1] \longrightarrow M$, with $\gamma(0) = p_0$ and $\gamma(1) = p_1$, under homotopy equivalence. The category $\mathcal{C}$ is known as the *fundamental groupoid* of M.

Definition 2.4.8 *If $\mathcal{C}$ and $\mathcal{C}'$ are two categories, then by a functor from $\mathcal{C}$ to $\mathcal{C}'$ is meant a map $F\colon \mathrm{Obj}\,(\mathcal{C}) \longrightarrow \mathrm{Obj}\,(\mathcal{C}')$, along with, for each pair of objects, $A, B \in \mathrm{Obj}\,(\mathcal{C})$, a map $F\colon \mathrm{Morph}\,(A, B) \longrightarrow \mathrm{Morph}\,(F(A), F(B))$ such that,*

(i) $F(\mathrm{id}_A) = \mathrm{id}_{F(A)}$;

(ii) $F(f \circ g) = F(f) \circ F(g)$ *for any compatible morphisms f and g in $\mathcal{C}$.*

One may now reformulate axioms A1–3′ in terms of the notion of a category, by stating that a $(d+1)$–dimensional topological field theory defines a functor $Z\colon \mathcal{M} \longrightarrow \mathcal{V}^1$, where $\mathcal{M}$ denotes the category whose objects are closed, oriented, d–dimensional manifolds and whose morphisms are cobordisms. The axioms also imply that this functor preserves various other structures which exist on both these categories, namely $\emptyset$ and $\amalg$ as well as * on $\mathrm{Obj}\,(\mathcal{M})$.

Domain categories

However, as mentioned immediately following Remark 2.2.1, it is actually necessary to take into account maps induced by isomorphisms between manifolds. Our aim is to reformulate the axioms of TFT so that they just require that Z be a map from an appropriate algebraic structure on the collection of 'all' manifolds, to a 'linearised' example of this structure, which preserves the whole structure. The required structure will take into account both the category structure in $\mathcal{M}$ and the existence of isomorphisms. To determine what structure we need, it is necessary to go back to the set of axioms for TFT in §2.2, and carefully observe all the operations on manifolds which were used.

Let $\mathcal{M}_0$ denote the collection of $(d+1)$–dimensional oriented manifolds with boundary and $\mathcal{M}_1$ denote the collection of d–dimensional oriented closed manifolds, where as usual the word 'manifold' is meant to include any additional structures and restrictions. The suffices are used to specify the codimension of the objects in these sets. Then isomorphisms between manifolds make $\mathcal{M}_0$ and $\mathcal{M}_1$ into categories. For $i = 0, 1$, there are also special (initial) objects $\emptyset \in \mathcal{M}_i$ as well as the operations $\amalg$ and *. Since the latter two operations are consistent with isomorphisms of manifolds, they really define functors,

$$\amalg\colon \mathcal{M}_i \times \mathcal{M}_i \longrightarrow \mathcal{M}_i\,,$$
$$^*\colon \mathcal{M}_1 \longrightarrow \mathcal{M}_1\,.$$

The most important connection between $\mathcal{M}_0$ and $\mathcal{M}_1$ is the boundary operation. As seen in §2.3, there is also another important relation between these sets, namely $\Sigma \longmapsto \Sigma \times I$. This gives two additional functors

$$\partial\colon \mathcal{M}_0 \longrightarrow \mathcal{M}_1\,,$$
$$\times I\colon \mathcal{M}_1 \longrightarrow \mathcal{M}_0\,.$$

The fact that $\partial(\Sigma\times I) = \Sigma\amalg\Sigma^*$ translates into the existence of a canonical morphism between these two objects in $\mathcal{M}_1$. On $\mathcal{M}_0$, in addition to the operation of $\amalg$, the other important construction entering the axioms is that of gluing together boundary components. For this we need a top-dimensional manifold M along with an identification of ∂M with $\Sigma \amalg \Sigma^* \amalg \Sigma'$, for some surfaces Σ, Σ', and the result will be a manifold $\cup_\Sigma M$ whose boundary may be identified with Σ'. When this statement is recast in our categorical notation, the data consists of an element $M \in \mathcal{M}_0$ along with a morphism $f \in \mathrm{Morph}_{\mathcal{M}_1}(\partial M, \Sigma_1 \amalg \Sigma \amalg \Sigma^*)$, while the result is the pair

$$\cup_\Sigma M \in \mathcal{M}_0\,,$$
$$\cup_\Sigma f \in \mathrm{Morph}_{\mathcal{M}_0}(\partial(\cup_\Sigma M), \Sigma')\,.$$

We now arrive at the appropriate formal structure on $(\mathcal{M}_0, \mathcal{M}_1)$; it was introduced by Quinn and called a *domain category.*[13] Just as in Example 2.4.3 a category was seen to be a pair of sets along with a collection of maps which must satisfy a number of conditions, this new algebraic structure consists of the pair of categories $(\mathcal{M}_0, \mathcal{M}_1)$, along with objects and maps $(\emptyset, \amalg, ^*, \partial, \times I, \cup_\Sigma)$, which are required to satisfy a fairly long list of properties. These properties are obtained by translating into categorical language, a generating set for the equalities satisfied by the objects and maps just outlined, when considered in their natural setting of manifolds. For a formal definition, the reader is referred to [Q], where there are also given a number of constructions of domain categories without any geometric connotations, coming from purely algebraical data. The domain category based on manifolds will be denoted by $\mathcal{M}$, its dependence on d being omitted from the notation as d is assumed to be fixed and throughout clear from the context.

Another example of a domain category comes from the category of vector spaces over say, $\mathbf{C}$, for definiteness. We will think of it as the canonical 'linear' example of a domain catgeory. It is based on the pair of categories $(\mathcal{V}_0, \mathcal{V}_1)$ defined as follows. Here $\mathcal{V}_0$ has as its set of objects the collection of all pairs $(V, \mathbf{v})$, where V is a vector space and $\mathbf{v} \in V$, while a morphism from $(V, \mathbf{v})$ to $(W, \mathbf{w})$ is an isomorphism $f: V \longrightarrow W$, for which $f(\mathbf{v}) = \mathbf{w}$. The category $\mathcal{V}_1$ has vector spaces as objects, while the only allowed morphisms are isomorphisms. To make $(\mathcal{V}_0, \mathcal{V}_1)$ into a domain category, $\mathcal{V}$, we now pick out appropriate special objects and functors:

- $\emptyset$: the object $\mathbf{C}$ in $\mathcal{V}_1$ and $(\mathbf{C}, 1)$ in $\mathcal{V}_0$;
- * : the functor[14] $\mathcal{V}_1 \longrightarrow \mathcal{V}_1$ acting on objects according to $V \longmapsto V^*$;
- $\amalg$: the functor $\mathcal{V}_i \times \mathcal{V}_i \longrightarrow \mathcal{V}_i$ defined on objects by $V \amalg W = V \otimes W$ for $i = 1$ and $(V, \mathbf{v}) \amalg (W, \mathbf{w}) \equiv (V \otimes W, \mathbf{v} \otimes \mathbf{w})$ for $i = 0$.

The ∂ functor takes $(V, \mathbf{v})$ to V while $\times I$ takes V to $(V \otimes V^*, \mathbf{e})$, where $\mathbf{e} \in V \otimes V^*$ is the vector defined by the identity map $V \longrightarrow V$. Finally, if $M = (V, \mathbf{v}) \in \mathcal{V}_0$ and $f: V \longrightarrow W_1 \otimes W_0 \otimes W_0^*$, then $\cup_{W_0} M = (W_1, \mathbf{w}) \in \mathcal{V}_0$ and $\cup_{W_0} f = \mathrm{id}$, where $\mathbf{w}$ is the image of $f(\mathbf{v}) \in W_1 \otimes W_0 \otimes W_0^*$ under the contraction of the first two factors.

We can now reformulate the definition of a TFT given in §2.2. Whenever we use the word *functor*, we mean a map between two algebraic structures which preserves all of the structure.

---

[13] To include axiom A2′ we would require also a * operation on $\mathcal{M}_0$. This gives rise to what we call a *-domain category structure.

[14] This example can be extended to give a * operation on $\mathcal{V}_0$ by endowing all vector spaces with inner product structures, which morphisms in $\mathcal{V}_i$ are required to preserve. Then one defines $(V,\mathbf{v})^* \equiv (V^*, \mathbf{v}^*)$.

Definition 2.4.9 *A $(d+1)$–dimensional topological field theory is a functor $Z\colon \mathcal{M} \longrightarrow \mathcal{V}$, where now $\mathcal{M}$ and $\mathcal{V}$ refer to the domain categories of manifolds and vector spaces discussed above.*[15]

In Definition 2.4.9, all the complications of the axioms of TFT have now been wound into the domain category structure. The dimension, $d+1$, of the theory, now only enters into the construction of $\mathcal{M}$. This formulation also allows for considerable generalisation; indeed one can consider a TFT over *any* domain category, not just one arising from manifolds of a certain dimension and this point of view is adopted in [Q]. One may say that a domain category embodies all of the geometry of manifolds which we are allowed to know about in TFT.

Bordism categories

In the above discussion of the domain category $\mathcal{M}$ of manifolds, it is not immediately apparent how to reconstruct from $\mathcal{M}$ the basic category structure on the collection of closed codimension–1 objects given by cobordisms. We now give this construction; it can be carried out quite abstractly on any domain category $\mathcal{M}$ to give a category, $\mathcal{C}$. The set of objects in $\mathcal{C}$ is defined to be $\mathrm{Obj}\,(\mathcal{M}_1)$. If $\Sigma_1, \Sigma_2 \in \mathcal{M}_1$ then

$$\mathrm{Morph}_{\mathcal{C}}(\Sigma_1,\Sigma_2) = \big\{(M,f) \bigm| M \in \mathcal{M}_0, f \in \mathrm{Morph}_{\mathcal{M}_1}(\partial M, \Sigma_1^* \amalg \Sigma_2)\big\}/\sim,$$

where the equivalence relation $\sim$ is defined by $(M,f) \sim (M',f')$ if, and only if, there exists $F \in \mathrm{Morph}_{\mathcal{M}_0}(M,M')$ such that $f = f' \circ \partial F$. The category $\mathcal{C}$ is known as the *bordism category* associated with $\mathcal{M}$ and will be denoted by $\mathrm{BORD}\,(\mathcal{M})$.

It is an easy exercise to check that the bordism category associated with the domain category of vector spaces, $\mathcal{V}$, is just the category of vector spaces, $\mathcal{V}^1$, of Example 2.4.5. From the functorial nature of the construction of a bordism category, it follows that if $Z\colon \mathcal{M} \longrightarrow \mathcal{V}$ is a TFT in the sense of Definition 2.4.9, it will also define a functor $Z\colon \mathcal{M} \longrightarrow \mathcal{V}^1$ of ordinary categories, a fact consistent with the discussion immediately following Definition 2.4.8.

2.5 Inverse limit constructions

The definition of TFT given in the last section, allows one to consider theories over domain categories in which the basic objects are manifolds with additional data. Adding further additional data usually makes the construction of TFTs easier. On the other hand, the theory then only produces an invariant of a manifold endowed with this structure, which may be a long way from what one would like to mean by a topological invariant.[16] The compromise often employed is to construct a TFT with additional data and then attempt to eliminate this data using an *inverse limit* construction.

---

[15] Replacing the domain categories $\mathcal{M}$, $\mathcal{V}$ by $*$–domain categories gives a $*$–TFT, that is, a TFT satisfying axiom A2$'$. See footnote 14.

[16] For example, in this generalised sense, conformal field theories are 2–dimensional TFTs where the additional data is a metric, given up to conformal equivalence. An axiomatic mathematical definition of CFT was given by Segal in [S1] (see also [S2]) and it is on these axioms that the first axioms for TQFTs [A2] were based.

Suppose that $\mathcal{M}$ and $\mathcal{M}'$ are two domain categories and that $r\colon \mathcal{M}' \longrightarrow \mathcal{M}$ is a functor. For technical reasons it is necessary to assume that r is surjective on the set of objects[17] along with a further condition which will be encountered later. Suppose Z' is a TFT over $\mathcal{M}'$; we wish to construct a TFT over $\mathcal{M}$ in such a way that the following diagram commutes.

$$\begin{array}{ccc} \mathcal{M}' & \xrightarrow{Z'} & \mathcal{V} \\ {\scriptstyle r}\big\downarrow & \nearrow_{Z} & \\ \mathcal{M} & & \end{array}$$

One may think of the objects in $\mathcal{M}'$ (that is, elements of $\mathrm{Obj}\,(\mathcal{M}'_i)$ for $i = 0, 1$) as being objects of $\mathcal{M}$ endowed with additional data, such as a triangulation or a metric. This is indicated by writing objects in $\mathcal{M}'$ as pairs (M, α), where M is an object in $\mathcal{M}$ and α represents the additional data. In this notation, $r(M, \alpha) = M$ and $Z'(M, \alpha) \in Z'(\partial M, \partial\alpha)$.

It is first necessary to assume[18] that $Z'(M, \alpha) = Z'(M, \beta)$ for any data α, β agreeing on the boundary, that is for which $\partial\alpha = \partial\beta$. Our aim is, for each Σ, to establish maps between the vector spaces $Z'(\Sigma, \alpha)$ for different α's, pick a vector space onto which all these vector spaces are naturally mapped and then to define a particular vector, in this vector space, associated with each M for which $\partial M = \Sigma$.

Fix an object $\Sigma \in \mathcal{M}_1$. There is a natural category structure, $\mathcal{C}$, that may be placed on the set, $r^{-1}(\Sigma) \subset \mathrm{Obj}\,(\mathcal{M}'_1)$, of possible extra data on Σ, namely a variant[19] of the restriction of the bordism category of $\mathcal{M}'$. Define a functor $F\colon \mathcal{C} \longrightarrow \mathcal{V}^1$ by

$$F(\Sigma, \alpha) = Z'(\Sigma, \alpha) \text{ on objects } (\Sigma, \alpha) \in r^{-1}(\Sigma), \text{ and}$$
$$F(M', f') = Z'(f')\big(Z'(M')\big) \text{ on morphisms}^{20}\ (M', f') \text{ in } \mathcal{C}.$$

The functorial nature of F may be rephrased by stating that for each morphism $\phi\colon \alpha \longrightarrow \beta$ in $\mathcal{C}$, there is defined a linear transformation $F(\phi)\colon Z'(\Sigma, \alpha) \longrightarrow Z'(\Sigma, \beta)$ compatible with composition.

---

[17] That is, we assume it to be surjective as a map $\mathrm{Obj}\,(\mathcal{M}'_i) \longrightarrow \mathrm{Obj}\,(\mathcal{M}_i)$ for $i=0,1$.

[18] More precisely, the assumption is that whenever $M'_1, M'_2 \in \mathcal{M}'_0$ are such that there are morphisms $f' \in \mathrm{Morph}_{\mathcal{M}'_1}(\partial M'_1, \partial M'_2)$ and $f \in \mathrm{Morph}_{\mathcal{M}_0}(r(M'_1), r(M'_2))$ for which $r(f') = \partial f$, one also has $Z'(M'_2) = Z'(f')Z'(M'_1)$ as elements of the common vector space $Z'(\partial M'_2)$.

[19] Recall from §2.4 that BORD $(\mathcal{M}')$ forms a category whose set of objects is $\mathrm{Obj}\,(\mathcal{M}'_1)$. The category $\mathcal{C}$ which we need is obtained from BORD $(\mathcal{M}')$ by restricting the set of objects to $r^{-1}(\Sigma)$ and restricting morphisms to those pairs (M', f') for which $r(f') = \partial f$ for some morphism $f \in \mathrm{Morph}_{\mathcal{M}_0}(r(M'), \Sigma \times I)$. The complications in this definition stem from the full domain category structure of $\mathcal{M}'$, which contains both isomorphisms amongst objects of fixed codimension and cobordisms involving both types of object. To get a basic feel for the constructions of this section, the reader may find it instructive to consider those domain categories $\mathcal{M}$ for which $\mathcal{M}_i = \mathrm{TRIV}(\mathrm{Obj}\,(\mathcal{M}_i))$ for $i=0,1$, thereby replacing the existence of isomorphisms by equalities. See Example 2.4.4.

[20] A few words of explanation are in order. Here $(M', f') \in \mathrm{Morph}_{\mathcal{C}}(\Sigma'_1, \Sigma'_2)$ for some objects $\Sigma'_i \in r^{-1}(\Sigma)$, which means that $M' \in \mathcal{M}'_0$ while $f' \in \mathrm{Morph}_{\mathcal{M}'_1}(\partial M', \Sigma'^*_1 \amalg \Sigma'_2)$. Since Z' is a functor at the level of domain categories, it gives rise to a vector $Z'(M') \in Z'(\partial M')$ along with a linear transformation $Z'(f')\colon Z'(\partial M') \longrightarrow Z'(\Sigma'_1)^* \otimes Z'(\Sigma'_2)$. Hence $Z'(f')(Z'(M'))$ is a vector in $Z'(\Sigma'_1)^* \otimes Z'(\Sigma'_2)$ and may be therefore viewed as a linear transformation $Z'(\Sigma'_1) \longrightarrow Z'(\Sigma'_2)$.

Using F, we now define a vector space $Z(\Sigma)$, independent of additional data by,

$$Z(\Sigma) = \Big\{ \big(\mathbf{v}: \alpha \longmapsto \mathbf{v}(\alpha)\big) \,\Big|\, \mathbf{v}(\alpha) \in Z'(\Sigma, \alpha) \ \forall \alpha\,, \\ F(\phi)\big(\mathbf{v}(\alpha)\big) = \mathbf{v}(\beta) \ \forall \phi: \alpha \longrightarrow \beta \text{ in } \mathcal{C} \Big\}\,,$$

the set of all flat sections. An element of $Z(\Sigma)$ is now a choice, for each possible additional structure α on Σ, of an element of $Z'(\Sigma, \alpha)$, in such a way that they are compatible with respect to F. From the functoriality of F it follows that for each possible additional data α_0 on Σ, there is an isomorphism

$$Z(\Sigma) \cong \big\{ \mathbf{v} \in Z'(\Sigma, \alpha_0) \;\big|\; F(\phi)\mathbf{v} = \mathbf{v} \ \forall \phi: \alpha_0 \longrightarrow \alpha_0 \text{ in } \mathcal{C} \big\}\,,$$

assuming that any two objects in $\mathcal{C}$ are connected by a morphism.

From the assumptions on Z', it follows that F has 'trivial holonomy', that is, $F(\phi)$ depends only on the source α and target β of ϕ, and so may be denoted $F_{\alpha,\beta}$.[21] Then there is an isomorphism,

$$Z(\Sigma) \cong \ker(F_{\alpha_0,\alpha_0} - \mathrm{id}) \subset Z'(\Sigma, \alpha_0)\,,$$

for each possible additional data α_0 on Σ.

To define Z on an object $M \in \mathcal{M}_0$ for which $\partial M = \Sigma$, pick a representative of $r^{-1}(M)$, say $(M, \theta) \in \mathcal{M}'_0$ and define $Z(M)$ to be the choice

$$\Big(\alpha \longmapsto F_{\partial\theta,\alpha}\big(Z'(M, \theta)\big)\Big)\,.$$

This completes the definition of the inverse limit theory, Z over $\mathcal{M}$.

An example satisfying the above constraints is that where the extra data is that of a triangulation on a d–dimensional manifold, Σ. It is known that one may pass between any two triangulations by a sequence of local moves on the simplices. The moves replace a local configuration of j simplices with $d+2-j$ simplices, by replacing the union of a proper subset of the faces of a $(d+1)$–simplex by the union of the complementary set of faces. Such a sequence of moves can be thought of as providing a (possibly singular) triangulation of $\Sigma \times [0,1]$ and therefore Z' provides a map between the vector spaces associated with the two different triangulations of Σ. The map obtained is just the image under the functor F constructed above, of an appropriate morphism. This process will be seen in action in §3.2.

3: Two dimensional TFT

In this section we shall discuss 2–dimensional TFT, illustrating how the axioms enable a complete classification of the possible theories in terms of simple algebraic structures. In §3.1, decompositions of manifolds into elementary pieces will be used to obtain this classifying structure. In §3.2, a family of TFTs will be constructed, using the inverse limit construction of §2.5, by starting with theories on triangulated manifolds. The examples so obtained, depend upon a choice of data which is similar to, but different from, that arising in the §3.1. Finally, the relation between the data required in §3.2, and the algebraic structure classifying the theory in §3.1 is obtained.

---

[21] It is well-defined for all α and β by the assumption that there is a morphism in $\mathcal{C}$ connecting any pair of objects. Equivalently, one may consider F as a functor $\mathcal{C} \longrightarrow \mathcal{V}^1$ where $\mathcal{C}=\mathrm{ALL}(r^{-1}(\Sigma))$; see Example 2.4.4.

3.1 Classification

Identification of domain category

According to Definition 2.4.9, before we can start to classify 2–dimensional topological field theories, we must understand the structure of the associated domain category, $\mathcal{M}$. Our aim is to construct purely topological invariants, and therefore we take the objects in $\mathcal{M}_i$ to be oriented manifolds given only up to topological equivalence. In particular, an element, Σ, of $\mathcal{M}_1$ should represent a manifold with codimension–1, that is, a disjoint union of oriented copies of the circle, S^1. More precisely, Σ is specified by an integer $N \in \mathbf{N} \cup \{0\}$ (the number of components of Σ), along with an identification of $\amalg^N S^1$ with Σ, given up to orientation-preserving diffeomorphisms. Equivalently, each component of Σ is labelled with an element of $\{1, \ldots, N\}$ and is given an orientation.

Notation 3.1.1 For $N \in \mathbf{N} \cup \{0\}$, let $[N]$ denote the set $\{1, \ldots, N\}$ when $N \neq 0$ and $\emptyset$ when $N = 0$. If M is a manifold, let $c(M)$ denote the number of connected components of M and $\chi(M)$ its Euler characteristic. If M is triangulated, let $n_d(M)$ denote the number of d–simplices in the triangulation. Also $\mathbf{R}^-$ will denote the set of non-positive real numbers.

For our purposes it is sufficient to work with the following domain category, $\mathcal{N}$, which is a slight variation of $\mathcal{M}$. The elements in $\mathcal{N}_1$ will consist of pairs (N, α), where $N \in \mathbf{N} \cup \{0\}$ and $\alpha \in \{+, -\}^N$, that is, $\alpha: [N] \longrightarrow \{+, -\}$. The set of morphisms between two objects (N, α) and (M, β) will be empty unless $M = N$ and then it will consist of those permutations $\sigma \in S_N$ for which $\beta_i = \alpha_{\sigma(i)}$. Let C_n denote the circle in $\mathbf{C}$ of radius $1/3$ and centre n. The elements of $\mathcal{N}_0$ are given by compact 2–dimensional oriented surfaces with boundary, M, embedded in $\mathbf{C} \times \mathbf{R}^-$ along with a choice of orientation of the boundary components, in such a way that $\partial M = \cup_{n \in [N]} C_n \times \{0\}$ as an unoriented manifold, where $N = c(\partial M)$. A morphism in $\mathcal{N}_0$ is just an orientation preserving homeomorphism. The full domain category structure on $\mathcal{N}$ is given as follows.

(i) The objects $\emptyset$ are $(0, \emptyset) \in \mathcal{N}_1$ and the empty manifold in $\mathcal{N}_0$.

(ii) The operation * is $(N, \alpha) \longmapsto (N, -\alpha)$ in $\mathcal{N}_1$ and the reversal of all orientations in $\mathcal{N}_0$.

(iii) The operation $\amalg$ is given by $(N, \alpha) \amalg (M, \beta) = (M + N, \alpha\beta)$ in $\mathcal{N}_1$, where $\alpha\beta$ denotes the sequence α followed by the sequence β. On $\mathcal{N}_0$, $M_1 \amalg M_2$ is given by the union of M_1 and M_2, in which M_2 is translated (and possibly also deformed so as to avoid self intersections), so that its boundary is the union of circles $C_i \times \{0\}$ for $i \in c(\partial M_1) + [c(\partial M_2)]$.

(iv) The boundary functor $\partial: \mathcal{N}_0 \longrightarrow \mathcal{N}_1$ takes an object $M \in \mathcal{N}_0$ to the pair $\big(c(\partial M), \alpha\big)$ where $\alpha(i)$ is $+$ or $-$ according as the orientation supplied on the ith boundary component matches or otherwise, the orientation induced on the boundary by the orientation on M.

(v) The functor $\times I$ takes (N, α) to a disjoint union of N deformed cylinders whose boundaries are $C_i \cup C_{N+i}$, for $i \in [N]$ and for which the orientations of the $2N$ boundary circles $C_1, \ldots, C_{2N}$, are in accordance with the signs $\alpha(1), \ldots, \alpha(N), -\alpha(1), \ldots, -\alpha(N)$.

(vi) Suppose that $M \in \mathcal{N}_0$ and $f \in \mathrm{Morph}_{\mathcal{N}_1}(\partial M, \Sigma_1 \amalg \Sigma \amalg \Sigma^*)$. Say, $\Sigma = (N, \alpha)$, $\Sigma_1 = (N_1, \alpha_1)$. Then $c(\partial M) = 2N + N_1$ and f is given by a permutation $\sigma \in S_{c(\partial M)}$. Re-embed M in such a way that $\sigma = 1$. Then the manifold $\cup_\Sigma M$ is defined by gluing onto M, N cylinders joining the ith and $(i+N)^{\mathrm{th}}$ boundary components, for each $i = N_1 + 1, \ldots, N_1 + N$.

Fundamental objects

Having now specified the domain category over which to consider topological field theories, we are in a position to consider such a theory, say Z. Denote the image under Z of the object $(1, +) \in \mathcal{N}_0$, consisting of a single positively oriented component, by V. By axioms A2 and A3, it follows that Z is now completely determined on $\mathcal{N}_1$ and is given by,

$$Z(N, \alpha) = \bigotimes_{i=1}^{N} (V_{\alpha_i}) \,,$$

where $V_+ \equiv V$ and $V_- \equiv V^*$. The morphisms in $\mathcal{N}_1$ provide an action of S_N on the groupoid of tensor products of N vector spaces $V_\pm$. We shall assume that this action is generated by the natural identification of tensor products in opposite orders, $W_1 \otimes W_2$ and $W_2 \otimes W_1$.[22] Any 2–dimensional compact orientable surface with boundary can be obtained by gluing together copies of a disc and a trinion along boundary circles. Thus any element of $\mathcal{N}_0$ may be obtained by appropriately gluing together copies of the three elementary pieces shown in Figure 4, namely a disc with positively oriented boundary, a trinion with all three boundary components positively oriented and a cylinder with its two boundary components oriented negatively, in order of their depiction from left to right.

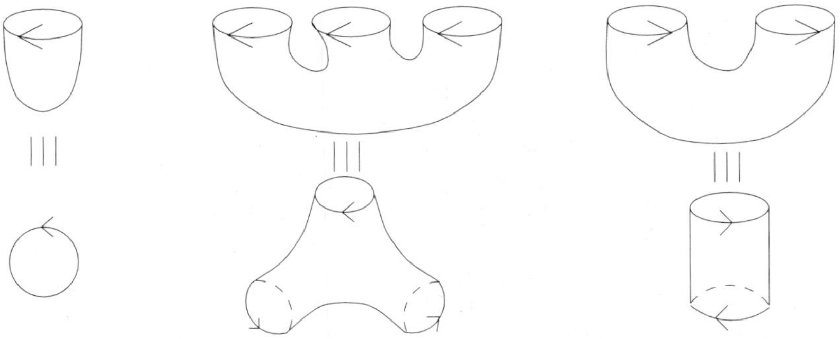

Figure 4: Elementary pieces for 2-d TFT

Under Z, these elementary surface pieces map to vectors which we denote by $1 \in V$, $t \in V \otimes V \otimes V$ and $\mu \in V^* \otimes V^*$, respectively. By axiom A0′, μ and t must be invariant under the actions of S_2 and S_3, respectively. By the gluing axioms A3 and A3′, it follows that it is possible to determine the image, under Z, of an arbitrary element of $\mathcal{N}_0$, as a contraction of a combination of 1, t and μ. Thus Z is determined by the collection $(V, 1, t, \mu)$.

[22] Allowing more general actions of S_N would provide the structure of a symmetric tensor category on the set of tensor products of copies of V and V^*.

Relations

Suppose that M is a compact oriented surface with boundary. Choose a collection of oriented simple closed curves, D_i, in Σ which are such that (the closures of) all components, T_j, of $M \setminus \cup_i D_i$ are homeomorphic to one of the forms in Figure 4. To such a decomposition one may correspond an oriented graph Γ with external legs, whose vertices are labelled by the components T_j and whose (internal) edges are in 1–1 correspondence with the curves D_i, while the external legs are labelled by connected components of ∂M. An edge labelled by D_i is oriented from T_j to T_k if the orientation on D_i matches that induced by the orientation of T_j (and consequently is opposite to that induced by the orientation of T_k). The vertices of Γ will have valencies 1,2 or 3 depending upon the type of the associated piece T_j, while at a vertex of valence v, all edges will be oriented outwards or inwards according as v is odd or even. This process is illustrated in Figure 5.

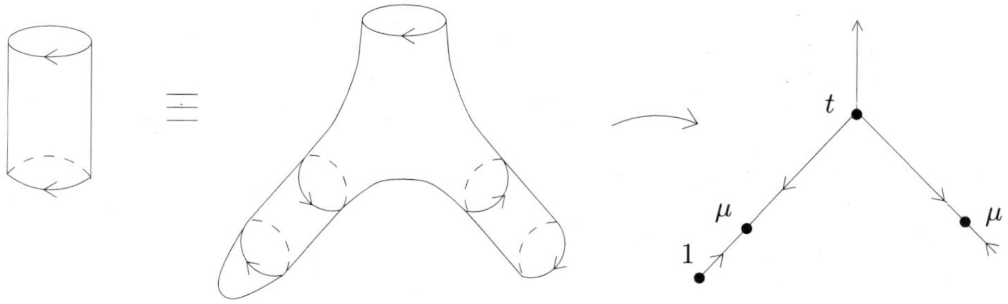

Figure 5: Decomposing a cylinder

Given such a decomposition, $Z(M)$ may be computed from Γ as follows. At each vertex, imagine placing copies of the vector spaces V and V^* on edges eminating from the vertex, according as the arrow on that edge points away from or towards the vertex, respectively. Next assign to each vertex one of the tensors 1, μ or t, according as its valence is 1, 2 or 3 respectively, where the factors of the tensor products of vector spaces to which these tensors belong are identified with the vector spaces already placed on edges. Since the tensors μ and t are totally symmetric, there is no ambiguity in this process. To obtain $Z(M)$, one contracts the (tensor) product of the tensors so placed at vertices, along all internal edges. For example, the contraction represented by the graph on the right of Figure 5 is the element of $V \otimes V^*$ given by $i_{S^1} = \mathrm{id}_V : V \longrightarrow V$ so that

$$(\mu \otimes \mathrm{id} \otimes \mu)(v \otimes t \otimes 1) = v\,, \qquad \text{for all } v \in V. \tag{3.1.2}$$

It is clear from the fact that only three types of vertices are allowed in Γ that Γ may be replaced by another graph, without loss of information, by removing all vertices of valence 2, along with all orientations on internal edges. One is left with an unoriented graph, whose external legs are oriented and all of whose vertices are of valence 1 or 3. To compute $Z(M)$ from this new graph, a copy of 1 or t is placed at each vertex according as it has valence 1 or 3, while it is understood that on internal edges the contraction $V \otimes V \longrightarrow \mathbf{C}$ is determined by μ.

For a fixed surface, M, there are many different possible decompositions while, by axiom A0, the results of the associated contractions of tensors must be equal. The equality of the images of the two decompositions of a sphere with 4 discs removed shown in the left hand side of Figure 6 yields the property,

$$\mu_{34}(t \otimes t) \in V^{\otimes 4} \text{ is invariant under the action of } S_4 \,. \tag{3.1.3}$$

Except for the sphere, disc and torus, any oriented surface can be obtained by gluing trinions without the use of discs, while any two such decompositions of a surface may be obtained from each other, up to homotopies of D_i, by a sequence of local moves of the two types shown in Figure 6. It follows that any quadruple $(V, 1, t, \mu)$ for which t and μ are totally symmetric, while properties (3.1.2) and (3.1.3) hold, may arise as the quadruple associated to a TFT. Such a quadruple forms the structure of a commutative *ambialgebra* in the terminology of [Q].

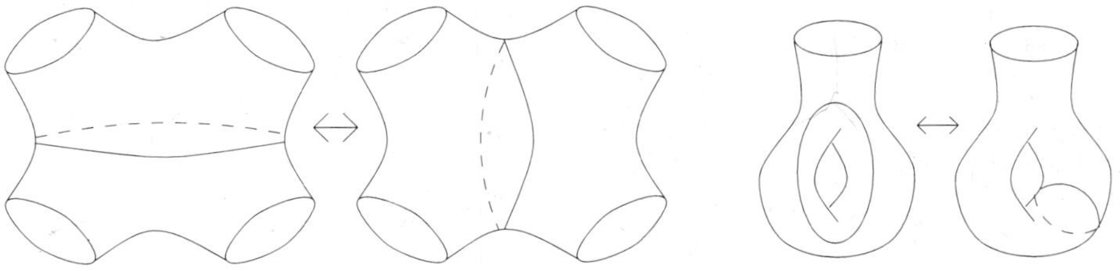

Figure 6: Moves on trinion decompositions

Complete solution

It follows from (3.1.2) that μ defines a non-degenerate inner product on V. Thus μ may be used to identify V with V^*. By contraction with two copies of μ, the element $t \in V \otimes V \otimes V$ gives rise to a map $m: V \otimes V \longrightarrow V$ which may be graphically represented by a trivalent vertex with two incoming arrows and one outgoing arrow. Considering m as a multiplication map, $a \otimes b \longmapsto ab$ makes V into a commutative associative unital algebra. (That $1 \in V$ is a unit, follows from (3.1.2); commutativity follows from the symmetry of t, while associativity comes from (3.1.3), which may be graphically depicted as in Figure 7.)

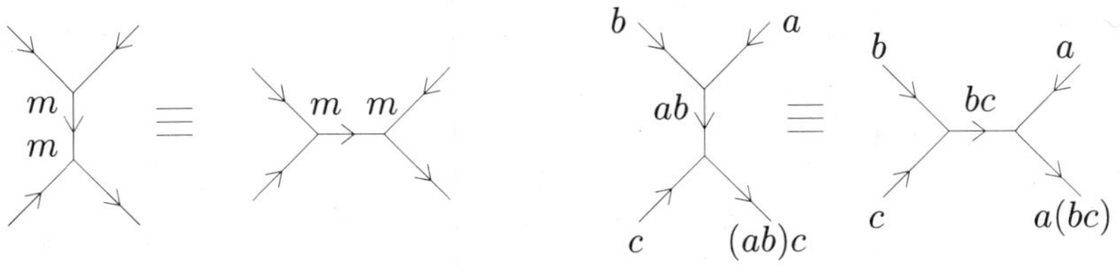

Figure 7: Associativity of m

Assume that V is finite-dimensional. By contracting both sides of (3.1.2) with $w \in V$, one may obtain $\mu(1, vw) = \mu(v, w)$ for all $v, w \in V$, so that the left multiplication operators on V are self-adjoint. It follows from this and the non-degeneracy of μ, that V contains no nilpotent elements and therefore that it is semi-simple. Hence there exists a basis for V consisting of idempotents $\{w_i\}$, say, for which $w_i w_j = 0$ whenever $i \neq j$. The unit here is $1 = \sum_i w_i$. Let $\lambda_i = \mu(w_i \otimes w_i)$. Then $t = \sum_i \lambda_i^{-2} w_i \otimes w_i \otimes w_i$. A surface M of genus g and with r boundary components can be decomposed into $2g - 2 + r$ trinions by $3g - 3 + r$ interior curves.[23] If ∂M has an orientation which on precisely s components matches that induced by the orientation of M, then $Z(M)$ is given in terms of the basis by,

$$Z(M) = \sum_i (\lambda_i^{-2})^{2g-2+r} \lambda_i^{3g-3+r} w_i^{\otimes r} = \sum_i \lambda_i^{1-g-r} w_i^{\otimes r} \in V^{\otimes r} \cong V^{\otimes s} \otimes (V^*)^{\otimes (r-s)}.$$

This completes the classification of 2–dimensional TFTs and the explicit evaluation of Z.

Remark 3.1.4 If the assumption that $i_{S^1} = \mathrm{id}_V$ is dropped, then the algebra structure on V need no longer be semisimple. However, by §2.3B, such an extension produces a larger class of theories which, in a suitable sense, contain no further information. The general commutative associative unital algebras arising from the general case are known as Frobenius algebras.

Remark 3.1.5 Assuming that V is a vector space on which a non-degenerate pairing, μ is given, one may ask from which commutative associative unital algebra structures on V, can one reconstruct a TFT, where one drops the requirement given at the end of §2.3B. The answer is that it is only necessary that the 3–tensor obtained from the multiplication operation by lowering one index using μ, be completely symmetric. Equivalently, the condition is that left muliplication is self-adjoint.[24]

3.2 Theories derived from triangulations

In this section we construct a particular example of a 2–dimensional TFT from algebraic data different from $(V, 1, t, \mu)$ of §3.1. This is done by first constructing a TFT over a domain category containing extra data and then applying the inverse limit construction of §2.5.

---

23 For this argument it is necessary to assume that $(g,r) \notin \{(0,0),(0,1),(1,0)\}$. However it may be separately checked that the result for $Z(M)$ holds in the exceptional cases. Alternatively, for any $u \geq 2$, M may be decomposed into $2g-2+r+u$ trinions and u discs by $3g-3+r+2u$ interior curves.

24 Let $\{e_i\}$ be a basis for V and denote the structure coefficients for the algebra on V by A_{ij}^k, so that $e_i e_j = A_{ij}^k e_k$, using the Einstein summation convention. Set $g_{ij} = \mu(e_i, e_j)$. Then the condition is that $A_{ij}^l g_{kl}$ be totally symmetric in i,j and k. This is ensured by $A_{ij}^l g_{kl} = A_{ik}^l g_{jl}$, or equivalently $\mu(e_i e_j, e_k) = \mu(e_j, e_i e_k)$.

Identification of domain category $\mathcal{N}'$

The extra data we will employ is that of a triangulation. Thus an object in $\mathcal{N}_1'$ will be a triangulated oriented 1–dimensional closed manifold, Σ. To reduce to combinatorial data, we insist that all our codimension–1 objects be rearranged so as to be unions, $\cup_{i\in[N]}C_i \times \{0\}$, of circles in standard position, as in §3.1. Such objects are specified by a pair (N, a) where $N \in \mathbf{N} \cup \{0\}$ and $a: [N] \longrightarrow \mathbf{Z}\backslash\{0\}$. Here $N = c(\Sigma)$ and $a(i)$ is an integer whose absolute value gives the number of vertices (or edges) in the triangulation of the ith boundary component, while the sign gives the orientation. The morphisms in $\mathcal{N}_1'$ are given by permutations of the components, there being no morphisms between objects (M, a) and (N, b) unless $M = N$ and $a = b \circ \sigma$ for some permutation $\sigma: [N] \longrightarrow [N]$.

The objects in $\mathcal{N}_0'$ will be triangulated, oriented, 2–dimensional manifolds with boundary, up to equivalences which preserve the boundary triangulation. More precisely, an object will be an embedding of a triangulated manifold in $\mathbf{C} \times \mathbf{R}^-$, whose boundary is of the form $\cup_{i\in[N]}C_i \times \{0\}$ where $N = c(\partial M)$, up to a change in interior triangulation. The morphisms in $\mathcal{N}_0'$ are topological equivalences of the associated manifolds, up to transformations which preserve the orientation of the manifold and the combinatorial structure of the boundary triangulation.

The rest of the domain category structure is much as for $\mathcal{N}$ and so we will not give it in detail except for noting that the operations $*$ and $\amalg$ on $\mathcal{N}_1'$ are given by,

$$(N, a)^* = (N, -a)\,, \qquad (N, a) \amalg (M, b) = (M + N, ab)\,,$$

where ab denotes the concatenation of the two sequences a and b.

The functor from the domain category $\mathcal{N}'$ to $\mathcal{N}$ is given by forgetting the additional triangulation data. In particular, objects in $\mathcal{N}_1'$ transform according to $(N, a) \longmapsto (N, \alpha)$ where $\alpha(i) = \text{sgn}(a(i))$.

Construction of TFT over $\mathcal{N}'$

Since a triangulated manifold is naturally expressed as the result of gluing[25] together copies of an elementary piece (a triangle), a natural way to construct a theory is to associate a tensor to a triangle and obtain $Z(M)$ by contraction along internal edges.

The data for the theory consists of a vector space, V, along with a tensor $t \in V \otimes V \otimes V$ and a pairing $\mu: V \otimes V \longrightarrow \mathbf{C}$. It is assumed that t is invariant under the action of $\mathbf{Z}_3$ by cyclic permutation while μ is symmetric. If Σ is an object in $\mathcal{N}_1$ given by the pair (N, a) then set $Z(\Sigma) = \bigotimes_{i=1}^{N} V_{\text{sgn}(a(i))}^{\otimes|a(i)|}$. Thus on a positively (negatively) oriented boundary component, there is a copy of V (V^*) assigned to each segment in the triangulation. Again, we make the symmetric group defining the morphisms act by permutation of the tensor product factors.[26]

[25] Observe that the form of gluing used here involves the identification of part-boundaries, namely along edges of the triangles, and is therefore more general than the gluing entering axiom A3′. Therefore one cannot conclude from the discussion following, that all TFTs can be obtained by the construction of this section, as one could in the case of decompositions into trinions in §3.1. However such a result would exist in the format of 2-ETFTs, see §5.

[26] To be absolutely correct, the morphisms in $\mathcal{N}_1'$ should not only allow permutations of the components but also rotations of the triangulation on each component, so that the correct group is $(\mathbf{Z}_{|a(1)|} \times \cdots \times \mathbf{Z}_{|a(N)|}) \rtimes S_N$ rather than S_N. We assume then that $\mathbf{Z}_a$ acts on $V^{\otimes a}$ by cyclic permutation of the factors.

If M is an object in $\mathcal{N}_0'$, we define $Z'(M)$ as follows. For simplicity, we assume first, that all the boundary components of M are positively oriented. Consider the tensor $t^{\otimes n_2(M)} \in V^{\otimes 3n_2(M)}$, see Notation 3.1.1. The factors[27] may be put in 1–1 correspondence with pairs (Δ_1, Δ_2) where Δ_1 is an edge in the boundary of the triangle Δ_2 in M. There are, therefore, precisely two factors associated with each internal edge and the result of applying μ on each such pair to $t^{\otimes n_2(M)}$ will be defined to be $Z'(M)$. Observe that this will be an element of $V^{\otimes x}$ where $x = 3n_2(M) - 2n_1^i(M) = n_1(\partial M)$ and $n_1^i(M)$ denotes the number of internal edges. If ∂M contains components which are negatively oriented then $Z'(M)$ is defined in the appropriate vector space by using the identification $V \cong V^*$ given by μ.

Since an object in $\mathcal{N}_0$ is only defined up to changes in interior triangulation, it is necessary to check that $Z'(M)$ is well-defined, that is, that it is independent of the choice of triangulation. Any two triangulations of M which agree on the boundary can be obtained from each other by a sequence of local moves, each being of one of the forms depicted in Figure 8. Thus $Z'(M)$ is well-defined so long as the local contractions of tensors remain unchanged under these two moves. Symbolically, this gives the following conditions on t and μ,

$$\begin{aligned} t = \mu_{24}\mu_{37}\mu_{68}(t \otimes t \otimes t) \in V^{\otimes 3}\,, \\ (\,\mathrm{id}^{\otimes 2} \otimes \mu \otimes \mathrm{id}^{\otimes 2})(t \otimes t) = (\,\mathrm{id}^{\otimes 4} \otimes \mu) P_{15}(t \otimes t) \in V^{\otimes 4}\,, \end{aligned} \tag{3.2.1}$$

where μ_{ij} denotes the action of μ on the ith and j^{th} copies of V and P_{ij} is the transposition of the ith and j^{th} factors. For any triple (V, t, μ) satisfying the above conditions, we have now constructed a TFT over $\mathcal{N}'$.

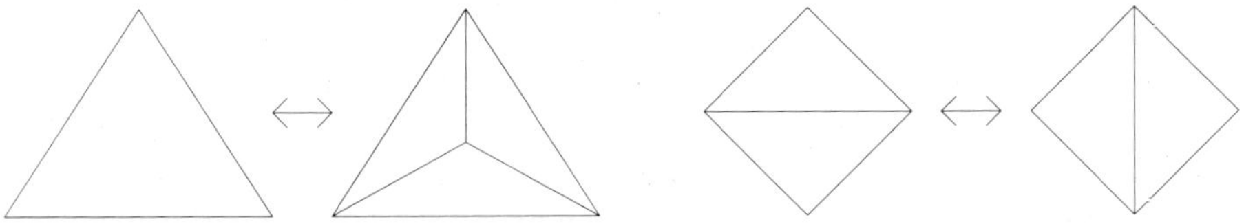

Figure 8: Local moves on triangulations

Graphical presentations

For any triangulated surface, M, the vector $Z'(M)$ has been defined as a contraction of tensors. As in §3.1, this may be pictorially represented by a graph, namely the graph, Γ, dual to the triangulation. All internal vertices of Γ have valence 3 (labelled by triangles in the triangulation), while the external legs correspond to the edges in the restriction of the triangulation to ∂M. The tensor t is represented by a trivalent vertex with all arrows directed outwards, while μ is represented by a bivalent vertex with arrows pointing inwards. To obtain $Z'(M)$, a copy of t is placed on each internal vertex of Γ, and of μ on each internal edge; external legs have a copy of μ attached if the orientation of the boundary component

[27] The order of the three factors associated with a triangle is chosen to be consistent with the cyclic order of the edges defined by the orientation on its boundary, induced by that on M. This precisely defines the tensor since t is $\mathbf{Z}_3$ invariant.

to which they belong, is negative. Since μ is symmetric while t is only invariant under cyclic permutation of the indices, such graphs must be drawn with attention to the cyclic ordering of the edges eminating from trivalent vertices. Once again, we omit bivalent vertices and internal arrows, since they can be deduced from the vertex types. The constraints given by Figure 8 are now represented by Figure 9.

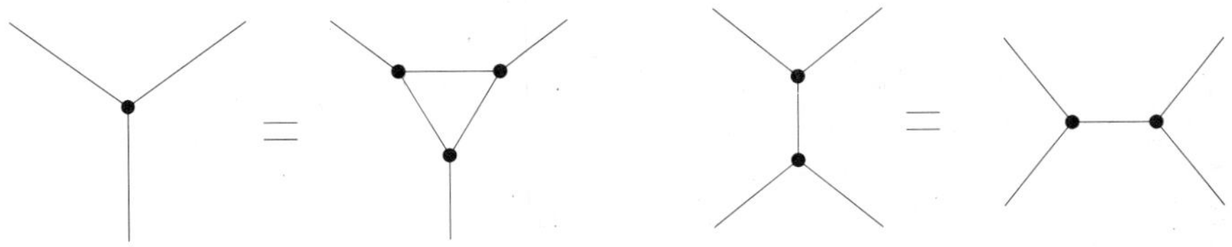

Figure 9: Diagrammatic formulation of relations on t and μ

As an example of the use of this notation, we show graphically that $F_{1,1} \equiv Z'(S^1 \times I)$ is an idempotent. Represent the cylinder by a triangulation in which there are two vertices and two triangles with precisely one edge in each boundary component, so that $F_{1,1}$ is presented pictorially by the left-hand side of Figure 10. Note that the notation $F_{1,1}$ is consistent with that used at the end of §2.5. The middle diagram in Figure 10 gives another representation of the cylinder, in which the two vertical edges marked with an arrow are identified. The right-hand part of Figure 10 gives the associated graph Γ, vertices representing traingles in the triangulation; the symbols a and b denote the two edges in the triangulation throughout.

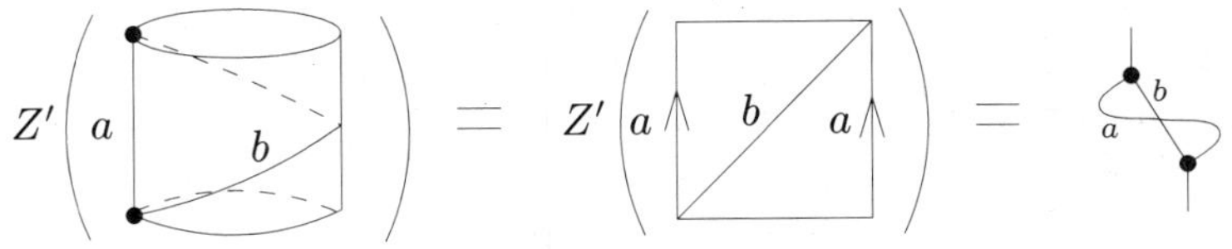

Figure 10: Z'(cylinder)

Place opposite orientations on the two boundary components, say positive on the upper component, so that $Z'(S^1 \times I)$ describes a map $V \longrightarrow V$ from bottom to top. Then $F_{1,1} \circ F_{1,1}$ is graphically represented by placing two copies of $F_{1,1}$ on top of each other; Figure 11 gives a graphical proof[28] of idempotency, using the relations in Figure 9.

The constraints on the triple (V, t, μ) in Figure 9 give rise to an algebraic structure much as in §3.1. Thus, by contracting t with one or two copies of μ, one may obtain maps $\Delta: V \longrightarrow V \otimes V$ and $m: V \otimes V \longrightarrow V$, see Figure 12.

[28] Each step in the proof involves either a regular isotopy rearrangement of the graph or an application of one of the two 'axioms' in Figure 9. Where the second axiom is used, a cross marks the central edge to which it is applied.

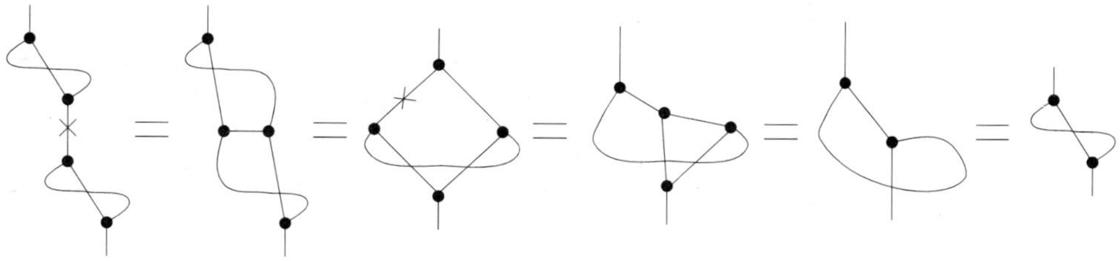

Figure 11: Verification that Z'(cylinder) is an idempotent

Figure 12: Construction of m and Δ

It follows from the right hand relation in Figure 9, that m is associative and Δ is coassociative, just as in Figure 7. Since t is only assumed to be invariant under the action of $\mathbf{Z}_3$, m need not be commutative. Also, there is no unit supplied. The two structures m and Δ are not necessarily compatible, in the sense that Δ may not define a homomorphism with respect to m, as can be seen from the example at the end of this section for general finite groups G.

Inverse limit construction

In order to obtain a TFT over $\mathcal{N}$ from Z', it is necessary to apply the inverse limit construction of §2.5. Since all closed codimension–1 objects can be written as disjoint unions of copies of S^1, the images of all such objects under Z are determined by $Z(S^1)$, with positive orientation. An object in $\mathcal{N}'$, representing a positively oriented circle, is specified by the number of segments into which the circle is subdivided. The image under Z' of a cylinder whose boundary components are oppositely oriented and are subdivided into m and n intervals, the former associated with the negatively oriented boundary component, will be a map $F_{n,m}: V^{\otimes m} \longrightarrow V^{\otimes n}$ and will be independent of the choice of interior triangulation. Picking a simple triangulation gives the presentation in Figure 13, all arrows on external legs pointing upwards. Since Z' defines a TFT over $\mathcal{N}'$, $F_{m,n} \circ F_{n,p} = F_{m,p}$ for all $m, n, p \in \mathbf{N}$. Indeed F is the functor of §2.5 when $\Sigma = S^1$ and $\mathcal{C}$ is the category whose objects are positive integers and whose morphisms are triangulated cylinders.

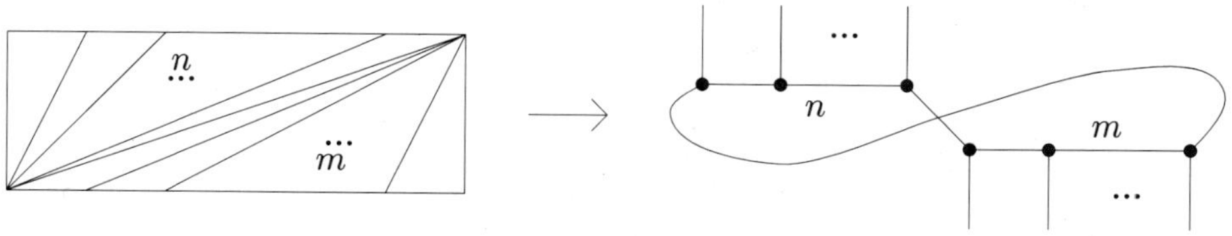

Figure 13: $F_{m,n}$

The vector space, $Z(S^1)$ associated with S^1, generated by the inverse limit construction, is therefore $\ker(F_{1,1} - 1) = \mathrm{Im}\,(F_{1,1}) \subset V$. Suppose that M is an arbitrary two-dimensional oriented surface in $\mathcal{N}_0$, whose boundary components are all positively oriented. Pick a triangulation, $\mathcal{T}$, of M. Then

$$Z(M) = \left(\bigotimes_r F_{n_r,1}\right) Z'(M, \mathcal{T}),$$

where n_r is the number of segments in the rth boundary component, in the restriction of $\mathcal{T}$ to ∂M.

Classification data

We have now constructed a TFT over $\mathcal{N}$, namely Z, from the data of (V, t, μ). By §3.1 any such TFT generates a quadruple $(V', 1, t', \mu')$ which describes, in particular, a unital commutative associative algebra. In our case, V' is the image of the map in Figure 10. The element $1 \in V'$ represents the image of a disc under Z. The simplest triangulation of a disc involves just one triangle, so that $1 = F_{3,1}\big(Z'(\text{triangle})\big)$ may be computed as in Figure 14.[29]

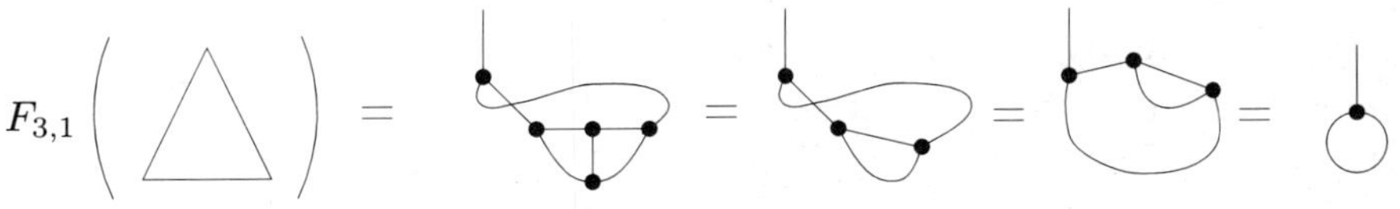

Figure 14: Unit in V'

The element t' comes from the image of a trinion. The simplest triangulation involves five triangles and gives the element of Figure 15.

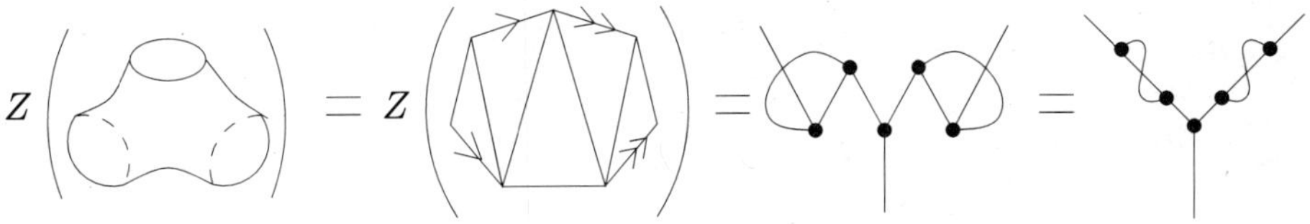

Figure 15: Image of trinion

It may be graphically verified, using manipulations similar to those of Figure 7, that t' defines a commutative algebra, as we know it must, by the analysis of §3.1.

29 In the second diagram, the lower vertex together with the three edges eminating from it represent Z'(triangle), while the rest of the diagram represents $F_{3,1}$ as given by Figure 13. In the rest of the figure, the axioms of Figure 9 are used to simplify the diagram.

An example

Suppose that G is a finite group and let V denote the vector space spanned by a basis $\{e_a | a \in G\}$, indexed by elements of G. Define t and μ by

$$t = |G|^{-2} \sum_{abc=1} e_a \otimes e_b \otimes e_c \,,$$
$$\mu(e_a \otimes e_b) = |G| \, \delta_{a=b^{-1}} \,,$$

where δ_X is defined to be 1 if X is true and 0 otherwise. By the last paragraph, the triple (V, t, μ) defines a TFT, say Z', over $\mathcal{N}'$. Suppose that $\mathcal{T}$ is a triangulation on a manifold M and assume that the orientations on all the boundary components are positive, while n_i is the number of segments in $\mathcal{T}$ on the ith boundary component. Then $Z'(M) \in V^{\otimes \Sigma n_i}$. Denote the set of simplices in $\mathcal{T}$ of dimension r by $\mathcal{T}^r$ and the set of edges in the restriction of $\mathcal{T}$ to ∂M by $\mathcal{T}^1_\partial$. Let $\mathcal{T}^1_{\text{or}}$ denote the set of pairs $\{(u,v) | u \in \mathcal{T}^1, v \in \mathcal{T}^2, u \subset v\}$; this may be thought of equivalently, as the set of oriented edges of $\mathcal{T}$, each internal edge appearing twice, while boundary edges appear only once.[30] Then, by construction,

$$Z'(M) = \sum_{p_\partial : \mathcal{T}^1_\partial \longrightarrow G} \Big(\bigotimes_{r \in \mathcal{T}^1_\partial} e_{p(r)} \Big) |G|^{|\mathcal{T}^1_i| - 2|\mathcal{T}^2|} \,.\, \begin{pmatrix} \# \text{ maps } p \colon \mathcal{T}^1_{\text{or}} \longrightarrow G \\ \text{compatible with } p_\partial \text{ and } \mathcal{T} \end{pmatrix} .$$

Here, by compatibility with $\mathcal{T}$, we mean that p assigns inverse elements to any two elements of $\mathcal{T}^1_{\text{or}}$ given by the same edge with opposite orientations, while the product of the elements associated with the three pairs (u,v), defined by a triangle v is $1 \in G$. Compatibility with p_∂ means that the restriction of p to the boundary is p_∂. This formula may be rewritten as,

$$Z'(M) = \sum_{p : \mathcal{T}^1_\partial \longrightarrow G} \Big(\bigotimes_{r \in \mathcal{T}^1_\partial} e_{p(r)} \Big) |G|^{|\mathcal{T}^2| - |\mathcal{T}^1| + |\mathcal{T}^0_i| - c(M) + c(\partial M)} \prod_{C_i} |[p_{C_i}]|^{-1}$$
$$\cdot \left(\# \text{ maps } \rho \colon \pi_1(M) \longrightarrow G \text{ such that } [\rho(C_i)] = [p_{C_i}] \ \forall i \right) ,$$

where $[a]$ denotes the conjugacy class of $a \in G$, C_i is the ith boundary component and $[p_{C_i}]$ denotes the conjugacy class of the product of the images under p of the segments in C_i; see Notation 3.1.1.

The maps $F_{n,1} \colon V^{\otimes n} \longrightarrow V$ are given by

$$e_{a_1} \otimes \cdots \otimes e_{a_n} \longmapsto f_{[a_1 \ldots a_n]} \,,$$

where $f_\alpha = |\alpha|^{-1} \sum_{a \in \alpha} e_a$. Hence the inverse limit construction generates $Z(S^1) = \langle f_\alpha \rangle$. The resulting theory, Z, has

$$Z(M) = |G|^{\chi(M) - c(M) - c(\partial M)} \sum_b \Big(\bigotimes_i f_{b(C_i)} \Big) \cdot \begin{pmatrix} \# \text{ maps } \rho \colon \pi_1(M) \longrightarrow G \\ \text{such that } [\rho(C_i)] = b(C_i) \ \forall i \end{pmatrix} .$$

[30] An element (u,v) of $\mathcal{T}^1_{\text{or}}$ is represented by the edge u with orientation given by the restriction of the orientation induced by v on its boundary, ∂v.

where the sum is over all maps b from the set of components of ∂M to the set of conjugacy classes in G. On closed surfaces, this reduces to $|G|^{\chi(M)-c(M)}$ times the number of homomorphisms $\pi_1(M) \longrightarrow G$.

The commutative unital associative algebra structure associated with this theory by §3.1, is a scaled version of the natural group algebra on G restricted to $\langle f_\alpha \rangle$, with unit $|G|.f_1$; that is, a scaled version of the centre of the group algebra of G. Equivalently, it has a basis whose coordinates with respect to $\{e_g\}$ are given by the characters, $\chi_i(g) \equiv \mathrm{tr}_{\rho_i}(g)$, of the irreducible representations, $\{\rho_i\}$, of G. Using the latter basis and the orthogonality of characters, it follows that $\{w_i \equiv (\dim \rho_i)\chi_i\}$ form a family of orthogonal idempotents, while $\lambda_i = (\dim \rho_i)^2$ in the notation of the last paragraph of §3.1.[31] By the calculations there, for a connected closed surface, M, of genus g, $Z(M) = \sum_i \lambda_i^{1-g}$, where the sum is over all irreducible representations ρ_i. Hence we obtain the equality,

$$\sum_i (\dim \rho_i)^{2-2g} = |G|^{1-2g} |\operatorname{Hom}(\pi_1(M), G)| ,$$

between the two different expressions for $Z(M)$, the left hand side obtained from the trinion decomposition and the right hand side from a triangulation. From a connected orientable surface of genus g with r boundary components, one obtains the more general equality, for all choices of conjugacy classes $b_1, \ldots, b_r$,

$$\sum_i (\dim \rho_i)^{2-2g-r} \prod_{j=1}^{r} (|b_j| \chi_i(b_j)) = |G|^{1-2g} \cdot \begin{pmatrix} \# \text{ maps } \rho\colon \pi_1(M) \longrightarrow G \\ \text{such that } [\rho(C_j)] = b_j \ \forall j \end{pmatrix} .$$

It is instructive to explicitly verify these equalities for small genus surfaces, for example for the sphere and torus, with and without one disc removed.

4: Category structures

The initial motivation for the axioms of §2 comes from physical ideas in field theory, where the image under Z of a cobordism, M from Σ_1 to Σ_2, is thought of as the operator defining the propagation across M of fields on Σ_1 to those on Σ_2. The slices Σ_i can be thought of as time-slices and M is the whole of space-time. In a topological field theory, there is no direction which is specially picked out as labelled by time, and indeed there is no particular reason why one should only consider decompositions by submanifolds of codimension 1.

Consider a topological field theory in dimension d. Suppose that M is a d–dimensional closed manifold and that it is split into two parts M_1 and M_2 by a $(d-1)$–dimensional manifold Σ. Then, as in (2.1.5),

$$Z(M) = \langle Z(M_1) \mid Z(M_2) \rangle ,$$

where $Z(M_1) \in Z(\Sigma)$ and $Z(M_2) \in Z(\Sigma)^*$. Hence, as was seen in §3.1, $Z(M)$ can be completely determined if it is possible to subdivide d–dimensional manifolds into elementary pieces and Z is known on each of these pieces. Suppose now that Σ is split into Σ_1 and Σ_2 by a $(d-2)$–dimensional manifold C. We would like to be able

[31] I thank A. Mann for bringing to my attention an error in a previous version of this computation.

to express $Z(\Sigma)$ as $\langle Z(\Sigma_1)|Z(\Sigma_2)\rangle$, where $Z(\Sigma_1)$ and $Z(\Sigma_2)$ are suitable objects in dual spaces $Z(C)$ and $Z(C)^*$, respectively. However, $Z(\Sigma)$ is to be a vector space, and so the question arises as to what sort of beast $Z(C)$ could be, so that an element of it can be paired with an element of a dual contraption to give a vector space. The answer we employ is that of a 2–vector space, and more generally, an n–vector space for manifolds of codimension n. This is our first motivation for the study of higher categories, since their definition naturally comes in the form of a structure on the collection of all n–vector spaces, analogous to the category structure on the collection of all vector spaces.

Just as domain categories were used in §2 to formalise the possible gluing operations appearing in TFTs, so more complex structures appear when more general gluing operations are allowed and provide a second motivation for the consideration of higher categories. Thus, allowing gluing down to manifolds of codimension 2, provides three types of objects, namely d–dimensional manifolds, $(d-1)$–dimensional manifolds and $(d-2)$-dimensional closed manifolds, forming sets $\mathcal{M}_0$, $\mathcal{M}_1$ and $\mathcal{M}_2$, respectively. There is now a particular object $\emptyset$ in each of these sets, while there are operations * on $\mathcal{M}_1$, $\mathcal{M}_2$ and $\amalg$ on all three sets. The sets are linked via two boundary functors, operations $\times I$ and numerous gluing maps, which allow gluing of objects at one codimension, along objects of higher codimension (up to a maximum of 2). Thus we allow only the operation of disjoint union on manifolds of codimension 2, while manifolds of codimension 1 may be glued along closed parts of their boundaries and top dimensional manifolds may be glued at corners.[32]

4.1 Higher category structures

As discussed in §2.4, a category may be viewed as a pair of sets along with various maps, namely source, target, identity and composition. The simplest way to extend a category structure[33] to higher dimensions, may be obtained by trying to fit it to a generalisation of Example 2.4.7, in which points and paths are generalised to embeddings of closed balls, D^p. Observe that such a closed ball can be decomposed into open cells,

$$A^p \cup A_0^{p-1} \cup A_1^{p-1} \cup \ldots \cup A_0^0 \cup A_1^0 \,, \tag{4.1.1}$$

where A^s denotes an embedding of a ball, B^s, whose boundary, ∂A^s, is given by the union of all A_i^t with $t < s$ and $i = 0, 1$.

The formalisation of the geometry just outlined, leads to the consideration of a structure known as a *spheric set*, see [MS]. Here there are objects of multiple types, forming sets $C^0, \ldots, C^n$, along with source, target and identity maps,

$$\begin{aligned} s_k, t_k\colon C^k &\longrightarrow C^{k-1}\,, \qquad (1 \le k \le n)\,; \\ i_k\colon C^k &\longrightarrow C^{k+1}\,, \qquad (0 \le k < n)\,. \end{aligned}$$

These maps must be compatible in the sense that the following equalities of compositions of maps hold, where, in each equality, k may range over all values for which one side of the equality is meaningful.

---

32 In this respect, we view disjoint union as a form of gluing along an empty object at the next higher codimension, although we also allow it on the top codimension.

33 See [Bn1] and [Bn2] for algebraic and topological motivations for higher algebra and category structures.

(a) $s_{k-1} \circ s_k = s_{k-1} \circ t_k$, $t_{k-1} \circ s_k = t_{k-1} \circ t_k$.

(b) $s_{k+1} \circ i_k = \mathrm{id}_{A_k} = t_{k+1} \circ i_k$.

Such a structure can be visualised in terms of a CW–complex, in which to each element of C^p there is associated a copy of D^p decomposed as a union of open cells, as in (4.1.1). The maps s and t take such an element to the two halves of the spherical boundary, namely

$$A_i^{p-1} \cup A_0^{p-2} \cup A_1^{p-2} \cup \ldots \cup A_0^0 \cup A_1^0 ,$$

with s being given by $i = 0$ and t by $i = 1$. Properties (a) and (b) allow one to construct maps $s_{p,q}, t_{p,q}: C^p \longrightarrow C^q$ whenever $n \geq p > q \geq 0$ by composing s and t maps; geometrically, the images of the object of (4.1.1) under these maps are the closures of A_0^q and A_1^q, respectively.

To get the notion of a higher category from that of a spheric set, it is necessary to require that there also be a collection of composition maps, $\mu_{p,q}$, defined for all $n \geq p > q \geq 0$,

$$\{(u, v) \in C^p \times C^p | s_{p,q}(v) = t_{p,q}(u)\} \longrightarrow C^p ,$$

which we denote by $(u, v) \longmapsto v \circ_q u$. The source of $v \circ_q u$ is given in terms of those of u and v by,

$$s_p(v \circ_q u) = \begin{cases} s_p(u), & \text{if } q = p - 1; \\ s_p(v) \circ_q s_p(u), & \text{if } q < p - 1, \end{cases}$$

and similarly for the target. The composition maps are also required to satisfy certain identity and associativity properties generalising those for standard categories.[34] The notion we arrive at, is known as that of a *small n–category.* Its full definition may be found in [MS].

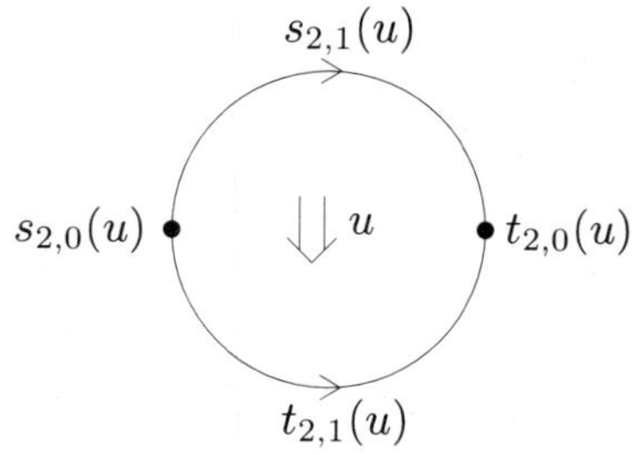

Figure 16: 2–objects in a 2–category

[34] There is another way of viewing the notation of a small n-category in terms of that of an ordinary category. View a category as a structure on two collections of objects, as in Example 2.4.3. Then an n-category gives rise to n collections of objects along with a category structure on each pair of such collections, the 'smallness' property being embodied in the fact that there are only two distinct compositions of source and target maps between any two of the collections of objects.

In accordance with the above geometrical imagery, compositions of type p objects whose source and target match at the level of type q objects, is represented by gluing along the common q-dimensional disc. Thus in a small 2-category, there are three object types, namely elements of C^0, C^1 and C^2 and the confusing terminology of, objects, 1-morphisms and 2-morphisms, respectively, is often adopted for elements of these sets. They are represented by points, arcs and 2-cells. See Figure 16 for the standard representation of an element of C^2. There are three possible compositions of objects in a 2-category, namely, $\mu_{1,0}$, $\mu_{2,0}$ and $\mu_{2,1}$ which are illustrated in Figure 17.

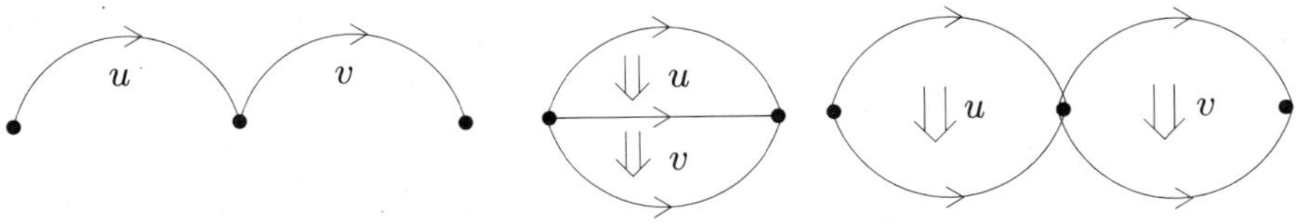

Figure 17: Compositions in a 2-category

Independence of the result of a method for evaluating the composition of a collection of compatible objects, upon the order in which the composition maps are applied, is ensured generally by the truth of this statement for the two particular arrangements of objects shown in Figure 18.

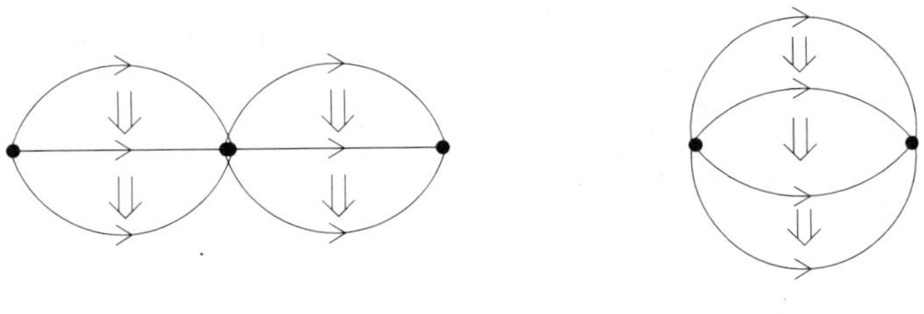

Figure 18: Associativity constraints in a 2-category

The higher category structure we have just constructed, is said to be *strict*, because it is required that the composition maps be absolutely associative. In weaker forms of category structures, it is not required that the equalities of compositions hold strictly, but merely that there are morphisms, given as part of the category structure, between any two such compositions. Thus for example, in a weak 2-category, it is not required that $\mu_{1,0}$ is strictly associative, but rather that, for each triple of compatible 1-objects, u, v and w, an invertible[35] 2-object $c_{u,v,w}$, known

[35] The notion of invertibility of an r-object, x, for $r>0$ is the formal analogue in higher categories of requiring a morphism to be an isomorphism in a usual category. Thus, it requires the existence of another r-object, y, whose source and target are those of x reversed and is such that the compositions $x\circ_{r-1}y$ and $y\circ_{r-1}x$ are both isomorphic to identity r-objects on appropriate $(r-1)$-objects. In the case of Example 2.4.6, the notion of invertibility of a morphism reduces precisely to the requirement that the associated map is literally an isomorphism.

as an *associativity constraint*, is given, whose source and target are $(w \circ_0 v) \circ_0 u$ and $w \circ_0 (v \circ_0 u)$, respectively. It is further required that this 2–object behaves 'naturally', in a suitable sense, under morphisms of u, v and w, and that given four compatible 1–objects u, v, w and t, a compatibility condition holds amongst the possible associativity constraints. This is illustrated in Figure 19, where the symbol '$\circ$' has been used to denote $\circ_0$. The structure now obtained is known as a *bicategory*, see [B].

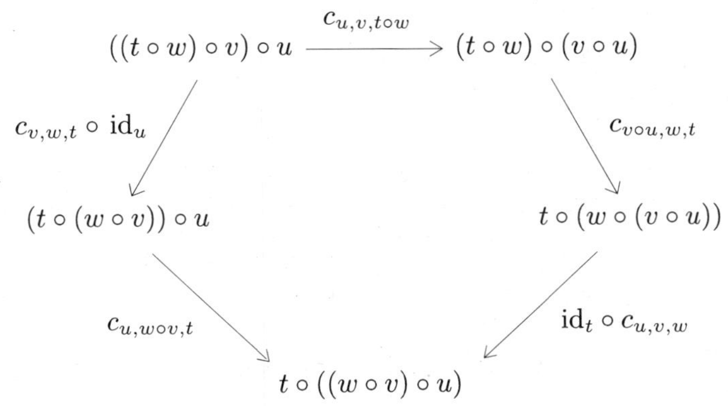

Figure 19: The pentagon relation in a bicategory

Although strict categories suffice[36] for the purposes of constructing higher vector spaces, it is a higher category in the weaker sense which is needed for the higher version of a domain category, just as it was necessary in a domain category to consider isomorphisms between manifolds rather than equalities. Unfortunately, although the notion of a usual category is perfectly well defined, there is at present no unique notion of a maximally weak higher category. A number of different notions of higher category, along with 'categorifications' of other algebraic structures, have been constructed piecemeal over the last 30 years. Although they all seem to fit into the same general framework, it is not currently possible to completely automate the process of categorification, some manual intervention being necessary in each case to decide precisely which of several possible slightly different structures is appropriate. It is, however, instructive to give the general idea of how, starting from the notion of *one* algebraic structure, another notion 'at the next level of categorification' may be constructed. (I believe this basic idea is due to Grothendieck.)

Suppose that $\mathcal{A}$ denotes a *formal* algebraic structure, that is, it is not a specific example, but rather it is a meta-object defining the notion of that particular type of structure. Hence it consists of a collection of sets, a collection of maps defined on sets generated by the basic sets and a collection of axioms which must be satisfied by these maps. Examples of such meta-objects include the notions of a set, groupoid,

36 Actually a complication arises with the identification of the order of the basis of a tensor product of three higher vector spaces.

group, ring and even of a category. The collection of axioms[37] may be thought of as a generating set with respect to our deduction system, for the set of all universally valid statements in a type $\mathcal{A}$ structure. The categorification process replaces types of object in the definition of $\mathcal{A}$ by other types, according to the list below.

A set A	$\longrightarrow$	a category A.
A map $f: A \to B$	$\longrightarrow$	a functor $f: A \to B$.
An element $x \in A$	$\longrightarrow$	an object $x \in A$.
An equality $a = b$ with proof P	$\longrightarrow$	an invertible morphism $c_P \in \mathrm{Morph}\,(a, b)$.
Concatenation of proofs	$\longrightarrow$	composition of morphisms.
Two different proofs P, Q of the same statement	$\longrightarrow$	equality of morphisms $c_P = c_Q$.

By a proof is meant a sequence of deductions, each step being an instance of one of the axioms of $\mathcal{A}$, or of the basic deduction system. The relations in the new structure thus arise as relations amongst relations in the old structure. The ambiguity in the new structure comes from the freedom available in deciding exactly what deductions are allowed in a proof, along with the fact that the above process may give inequivalent outputs starting from equivalent initial structures.[38] By this type of procedure, category structures have been generated from many different algebraic structures.

set	$\longrightarrow$	category
groupoid	$\longrightarrow$	monoidal category
abelian groupoid	$\longrightarrow$	braided monoidal category
group	$\longrightarrow$	rigid monoidal category
ring	$\longrightarrow$	ring category [39]
module over a ring	$\longrightarrow$	module category over a ring category
category	$\longrightarrow$	bicategory
braided monoidal category	$\longrightarrow$	braided monoidal 2–category [40]
bicategory	$\longrightarrow$	tricategory [41]
Hopf algebra	$\longrightarrow$	Hopf category [42]

If the constraints defining the initial algebraic structure are described by means of the commutativity of 2–faces of a polyhedron, P, then the constaints defining its categorification may be described as the commutativity of 3–faces of P, or as the commutativity of 2–faces of an appropriate complex Q. If P is considered as a

[37] Since the maps in the structure are often between sets obtained from the basic ones using Cartesian product and other set operations, one should adjoin to the axioms supplied for $\mathcal{A}$ also axioms of set theory.

[38] For example, starting with the notion of a category, as given in Definition 2.4.1, the process generates the notion of a bicategory. However, if we choose to define a category according to Example 2.4.3 then the resulting structure will be slightly different, since a composition will be defined whenever there are given objects $u, v \in C^1$ and a morphism $f \in \mathrm{Morph}_{C^0}(s(u), t(v))$, rather than only when the source of one object matches the target of the other.

[39] See [K], [Lap].

[40] Here there is some ambiguity as to the 'right' definition. See [Br] and [KV2] for the best lists currently known.

[41] See [GPS].

[42] See [CF].

ranked poset and has unique minimal and maximal elements p_0 and p_1, respectively, then Q is a complex, whose vertices label the maximal chains from p_0 to p_1 in P. For an analysis of the form of Q when P is a permutahedron (so that the constraints on the initial algebraic structure are described by the quantum Yang-Baxter equation), the reader is referred to [L3].

4.2 Higher vector spaces

Just as ordinary vector spaces can be considered as objects in an appropriate category, $\mathcal{V}^1$, higher vector spaces appear as objects in an appropriate higher category[43] $\mathcal{V}^n$. To illustrate the structure, we will discuss the case of 2–vector spaces only, as introduced in [KV2]. Formally, one should think of a 2–vector space as a linear space over the category of vector spaces, just as an ordinary vector space is a linear space over the field of complex numbers. Everywhere that $\mathbf{C}$ previously entered, it is replaced by $\mathcal{V}^1$, while the operations of addition and multiplication in $\mathbf{C}$ are replaced by direct summation and tensor product of vector spaces.[44]

Up to isomorphism, a 2–vector space is specified by its rank, a non-negative integer.[45] Thus, if $\mathcal{V}$ is a 2–vector space of rank n, then an identification with the standard 2–vector space of rank n is equivalent to a choice of basis, $\{\mathbf{e}_1, \ldots, \mathbf{e}_n\}$ for $\mathcal{V}$. With respect to this basis, any element, V, of $\mathcal{V}$ is specified by its coordinates, a list of n vector spaces, $(V_1, \ldots, V_n)$, and this fact may be formally denoted by,

$$V = \bigoplus_{i=1}^{n} V_i \cdot \mathbf{e}_n \ ,$$

where '$\cdot$' is the analogue of scalar multiplication. If $\mathcal{V}$ and $\mathcal{W}$ are two 2–vector spaces, of ranks n and m respectively, then a morphism, S, between them should be a "linear transformation". In terms of bases it is specified by an $m \times n$ matrix of vector spaces, $S_{i,j}$, such that if $S(V_1, \ldots, V_n) = (W_1, \ldots, W_m)$ then

$$W_i = \bigoplus_{j=1}^{n} S_{i,j} \otimes V_j \ .$$

Two morphisms S and T from $\mathcal{V}$ to $\mathcal{W}$ cannot be considered as equal, as it is only possible to talk about (2–) morphisms between them. Such a 2–morphism would be given by a family of linear transformations $S_{i,j} \longrightarrow T_{i,j}$ indexed by the pair (i, j). That is, in terms of coordinates, a 2–morphism is given by a matrix of matrices.

To make the form of the 2–category structure, $\mathcal{V}_c^2$, on the collection of totally coordinatised 2–vector spaces clearer, we now give a purely combinatorial list of the sets of objects. A 0–object will be an element $n^0 \in \mathbf{N} \cup \{0\}$. A 1–object will consist of a triple, (n_0^0, n_1^0, n^1) where $n_i^0 \in \mathbf{N} \cup \{0\}$ $(i = 0, 1)$ and n^1 is an $n_1^0 \times n_0^0$ matrix

---

43 The higher dimensional analogue of the category $\mathcal{V}^1$ of uncoordinatised vector spaces, should be a higher category in the as yet undefined (except for a 2–category) weak sense. To avoid this problem, we only discuss the analogue of $\mathcal{V}_c^1$ in this section. We also take $\mathbf{C}$ as the ground field.

44 It is actually possible to make this more precise by defining a 2–vector space to be a module category over the ring category of vector spaces. See [KV2].

45 In categorical language this reads: up to module equivalence, a 2-vector space may be identified with the module category $(\mathcal{V}^1)^n$, of rank n, over $\mathcal{V}^1$, for some non-negative integer n.

whose entries are elements of $\mathbf{N} \cup \{0\}$.[46] A 2–object will consist of five pieces of data

$$(n_0^0, n_1^0, n_0^1, n_1^1, n^2)\ ,$$

where n_i^0 are non-negative integers, n_i^1 are $n_1^0 \times n_0^0$ matrices with non-negative integer entries and n^2 is an $n_1^0 \times n_0^0$ matrix, whose $(i,j)^{\text{th}}$ entry is an $n_1^1(i,j) \times n_0^1(i,j)$ matrix with complex entries. The source and target maps are given by forgetting parts of the information, while the composition operations are generalisations of matrix multiplication.

This definition can be generalised to totally coordinatised n–vector spaces without any major changes. However, as was mentioned above, in real applications, what is wanted is a category of uncoordinatised higher vector spaces, which must necessarily form only a weak higher category.

5: Extended Topological Field Theories

The axioms of §2 were formulated by considering the properties that would be expected of Feynman integrals over manifolds with boundary, under various gluing operations. As noted at the start of §4, if objects are to be associated with codimension-1 manifolds with boundary, in such a way that they behave reasonably under gluing, then these objects must be elements of a 2–vector space.

Let us fix the top dimension, d, of the theories under consideration (that is, the dimension at which manifolds are assigned vectors). By an r–ETFT (*extended topological field theory*) we mean an assignment of objects to manifolds of codimension up to r, which behaves naturally under gluing operations. This assignment will give,

codimension k closed manifold, Σ	$\longrightarrow$	k–vector space $\overline{Z}_k(M)$, for $0 \leq k < r$;
codimension k arbitrary manifold, M	$\longrightarrow$	element $Z_k(M)$ of the $(k+1)$–vector space $\overline{Z}_{k+1}(\partial M)$, for $0 \leq k \leq r$.

The only gluing operations allowed are those in which all parts of the gluing are expressed in terms of manifolds of codimension at most r, while those manifolds which appear at codimension r are all closed.

Notice that in §4.2, the collection of k–vector spaces was considered as forming a k–category, $\mathcal{V}^k$, while a particular k–vector space was considered as a linear space over $\mathcal{V}^{k-1}$ in place of $\mathbf{C}$. An element of a coordinatised k–vector space of rank m is specified by a list of $(k-1)$–vector spaces of length m, namely, the *coordinates* of the element with respect to a basis. This statement at $k = 2$ translates into the fact that a 2–vector space is a (1–) category and that, with respect to a basis, an element of a 2–vector space is given by a sequence of vector spaces. Pushing down to $k = 1$, we see that a 1–vector space is an ordinary vector space and, with respect to a basis, an element is specified by a sequence of complex numbers. In this sense

[46] This is because, up to isomorphism, a vector space is specified by its dimension, so that a matrix of vector spaces turns into a matrix of dimensions.

we think of a 0–vector space as being a complex number, while the collection of 0–vector spaces forms the set (= 0–category) $\mathbf{C}$.

When $r = 1$, the structure of an r–ETFT therefore associates complex numbers and vector spaces to closed codimension–0 and 1 manifolds, while associating vectors to arbitrary codimension 1–manifolds; thus a 1–ETFT is just an ordinary TFT. For ordinary TFTs, these three structures are not all independent, in the sense that the vector associated to a codimension–1 manifold, M, in the case when M is closed, is an element of $Z(\partial M) = Z(\emptyset) = \mathbf{C}$. In a similar way, the structures associated with arbitrary manifolds and with closed manifolds are identified, by the requirement that $\overline{Z}_k(\emptyset) = \mathbf{1}_k$. Here $\mathbf{1}_k$ is the one–dimensional k–vector space, that is, it has rank one over $\mathcal{V}^{k-1}$. Thus, if M is a closed codimension k manifold, then $Z_k(M) \in \mathbf{1}_{k+1}$, so that it is a k–vector space, and we require that it be precisely $\overline{Z}_k(M)$.

In principle, it is fairly straightforward to define the notion of an r*–domain category* to be the generalisation of the notion of a domain category (for $r = 1$), which includes all the object types above and the allowed gluing operations which may be performed upon them. A d–dimensional r–ETFT would then just be a functor $\mathcal{M}^r \longrightarrow \mathcal{V}^r$, where $\mathcal{M}^r$ is the r–domain category of manifolds of codimension up to r, with base dimension d, and $\mathcal{V}^r$ is the r–domain category of r–vector spaces. By an ETFT we will mean a d–ETFT, that is, it will assign structures to submanifolds all the way down to a point. Since, when gluing at arbitrary corners is allowed it is possible to subdivide any manifold into fixed basic pieces, namely simplices, an ETFT can be expressed in terms of the image of a simplex and the gluing rules. This approach shifts all the complication into the gluing rules, since, unlike the case of TFTs, the gluing operations may now involve corners and it is important to know the local structure of the manifolds being glued near such corners.

If $\mathcal{T}$ is a triangulation of a manifold M, then a subdivision which shows all the complexity of the gluing operations in the neighbourhood of subsimplices of $\mathcal{T}$ is obtained by a construction known as the *dual barycentric subdivision.* It produces a subdivision $\mathcal{T}^*$, for which there are 1–1 correspondences,

$$\begin{aligned} \text{top dimensional cells} &\longleftrightarrow \text{simplices and subsimplices,} \\ \text{vertices} &\longleftrightarrow \text{complete flags of simplices,} \\ \text{faces} &\longleftrightarrow \text{partial flags of simplices,} \end{aligned}$$ [47]

in which the left hand column represents structures in $\mathcal{T}^*$ and the right hand column is relevant to $\mathcal{T}$. For each subsimplex, Δ, in $\mathcal{T}$, the geometry of the associated top dimensional cell in $\mathcal{T}^*$ precisely specifies the complexity of the gluing operation required in the neighbourhood of Δ. By decomposing these cells into a small number of elemental pieces, it is possible to specify an ETFT by giving the structures associated with only a finite number of geometric types of cell. See [L2].

Writing out all of the gluing data in terms of bases, one obtains the following coordinatised description of an ETFT. It consists of a large quantity of data, sets and numbers. It is simplest to think in terms of cells that may arise in the decomposition

[47] A flag of simplices is a sequence $\Delta_0 \subset \Delta_1 \subset \cdots \subset \Delta_k$, for which each Δ_i is a subsimplex of the complex $\mathcal{T}$. A flag is called complete if $k = \dim M$ and $\dim(\Delta_i) = i$ for all i. The type of a flag refers to the list of dimensions of the simplices involved.

$\mathcal{T}^*$, for a generic $\mathcal{T}$, since these describe all the required gluing operations. The basic set-data is recursively given as a set of labels that may be assigned to vertices, along with, for each geometric type of face, of dimension $k < d$, a set of labels depending on a prior choice of labels on all the facets of the face. Finally, for each possible d–dimensional cell type, there is a weight supplied by the data, whenever all its subcells have been labelled by elements of the appropriate labelling sets. From such data, the scalar assigned to a triangulated top dimensional manifold is,

$$Z(M) = \sum_{\substack{\text{labellings}\\ \sigma}} \prod_{\substack{\text{subsimplices}\\ \Delta \in \mathcal{T}}} (\text{weight assigned to cell labelled } \Delta \text{ in } \mathcal{T}^*)\,,$$

where σ ranges over all allowed labellings of the subcells of $\mathcal{T}^*$. This expression may be viewed as a discretised form of a Feynman integral.

Example 5.1 The example of a TFT constructed in §3.2 may be viewed in the framework of ETFTs with $d = 2$ as follows. To vertices of $\mathcal{T}^*$, there is associated a labelling set with only a single element. To edges, there is associated a labelling set with 1 or $\dim V$ elements, depending upon the type. The weights on 2–dimensional cells are the matrix elements of t or μ, according to the cell type. In order for a set of combinatorial data to define an ETFT, certain conditions must be satisfied which guarantee the independence of the result upon the choice of triangulation. These conditions are generated by those coming from the local moves on triangulations in Figure 8, just as for the conditions on t and μ in §3.2.

Fortunately, for arbitrary d, there is a simple set of local moves on triangulations which generates the equivalence (see [Ma], [P]) and therefore the constraints on the data may be written down as a finite set of conditions similar to (3.2.1). It is also possible to recover a full topologically invariant ETFT by using an inverse limit construction similar to that employed in §3.2. For more details of small dimensional ETFTs, the reader is referred to [L2].

6: Conclusion

What we have indicated in these notes is how complicated algebraic and categorical structures can arise out of the consideration of gluing rules in higher dimensions. The study of topological field theory is really the study of space, and to this extent, any structure appearing in a classification of TFTs may be thought of as dual to an appropriate structure on higher dimensional geometric objects.

We have constructed our algebraic structures in a fairly naïve way, throwing in all the structure that must be present and then constraining it by the obvious conditions. The artful part comes next, namely extracting that part of the structure which is essentially interesting. Thus, just as in §3.1 it was seen that the essential algebraic structure in 2–dimensional TFT was that of a commutative unital associative algebra, in 3–dimensional TFT, it is a quasi-Hopf algebra.[48]

The relationship between algebraic and geometric structures is nicely illustrated with reference to the Yang–Baxter equation. It plays a central role in the theory

[48] See [F] for a differential geometric approach to the construction of the quasi-Hopf algebra associated to a TFT.

of quantum groups and so one would expect to be able to realise its geometrical significance. Indeed, one can obtain any compact connected 3–manifold by surgery around suitable links in S^3. Any link can be represented as the closure of an appropriate braid, and braids on a given number of strands form a group whose main relation is a variant on the Yang–Baxter equation. In fact, historically, it is this latter connection which came first, see [RT].

In the search for the construction of interesting higher dimensional TFTs, this mirroring of the geometry in algebra has been exploited to try to guess at appropriate algebraic structures, though the 'right' one has yet to be found. The simplest higher dimensional geometric objects are simplices and hypercubes, and various approaches at algebraic and categorical constructions to encapsulate their combination along codimension–1 faces have been investigated.[49] There have been various attempts at the construction of higher dimensional versions of the Yang–Baxter equation and braid groups, some of which arise both in connection with physics and the geometry of knots in higher dimensions.[50] Using purely categorical constructions, it is possible to construct from a suitably constrained braided monoidal tensor category, a four–dimensional TFT, but unfortunately the invariants it defines are expressible in terms of simple classical invariants of manifold such as the signature.[51] There have also been attempts, using the general language of §4.1, to construct category structures relevant in four dimensions, from those known to be relevant in three dimensions, such as [KV2] for braided monoidal 2–categories and [CF] for Hopf categories.

Even apart from the debate over which algebraic structures are natural in the study of higher dimensional topological field theories, there are many open questions remaining. For example, is it possible to extend all TFTs to ETFTs; that is, given a TFT, is it possible to reconstruct compatible data associated with objects of codimension greater than 1? Even if the answer to this question is "no", it is still likely that all 'interesting' TFTs (for example those coming from physics via the use of Feynman integrals) *do* arise from ETFTs. It is then interesting to try, for particular known theories, to compute the relevant set- and tensor-data structures which determine an ETFT; this is equivalent to supplying a state model on a triangulation whose partition function generates the invariants defined by the theory. Even for Jones-Witten theory this has not yet been carried out, although for 3-manifolds [TV] produces a state-sum presentation of the square of the modulus of the Witten-Reshetikhin-Turaev invariant. In four dimensions, it seems that Donaldson–Floer theory satisfies the TFT axioms and even has interesting structure at codimensions greater than one. Again, one may ask for a state-sum formulation for Donaldson invariants and perhaps the recent simplifications in this theory due to the invention of the Seiberg–Witten equation will help in this regard.

---

[49] See for example [KV1], [L1], [L4] and [St].

[50] See [Z1], [Z2] for the physical background behind the Zamolodchikov tetrahedron equation. In [MS], a family of higher braid groups is constructed. In [KV2], [L1] notions of a braided monoidal 2–category and a 3–algebra, respectively, are introduced and related to different forms of generalisation of the Yang–Baxter equation. See [CS1], [Fi], [CS2] and [KT] for category structures arising from 2–knots in 4–dimensions.

[51] See [GK], [Ku], [T] and [Y] for purely categorical constructions of 3–manifold invariants. Using similar methods, [CY] and [CKY] construct 4–manifold invariants, while [R] gives the basic proof that these invariants are 'trivial'.

Acknowledgements The author would like to thank all those who have made suggestions and comments on earlier versions of these notes, including A. Adler, L. Breen, P. Cartier, V. Kac, Y. Kosmann-Schwarzbach, A. Mann and I. Todorov. These notes, initially constructed as preparatory reading for the AMS Short Course 'Knots and Physics', are also based on the first two lectures of a series of four given at IHES in October/November 1994 and entitled *Four Introductory Lectures on Topological Field Theory.* The author would like to thank D. Ruelle and M. Gromov for inviting her to give these lectures and the IHES for their generous hospitality during the writing of these notes and for support from a Raymond and Beverly Sackler Fellowship.

References

[A1] M.F. Atiyah, 'The Geometry and Physics of Knots', *Lezione Lincee* Cambridge University Press (1990).

[A2] M.F. Atiyah, 'Topological quantum field theory', *Publ. Math. IHES* **68** (1989) p.175–186.

[B] J. Bénabou, 'Introduction to bicategories', *Lect. Notes in Math.* **47** (1967) p.1–77.

[Br] L. Breen, 'On the classification of 2-gerbes and 2-stacks', *Asterisque* **225** (1994) p.1–160.

[Bn1] R. Brown, 'From groups to groupoids: a brief survey', *Bull. London Math. Soc.* **19** (1987) p.113–134.

[Bn2] R.Brown, 'Symmetry, groupoids and higher-dimensional analogues', *Computers Math. Applic.* **17** (1989) p.49–57.

[CS1] J.S. Carter, M. Saito, 'Reidemeister moves for surface isotopies and their interpretation as moves to movies', *J. Knot Th. Ramif.* **2** (1993) p.251–284.

[CS2] J.S. Carter, M. Saito, 'Knotted Surfaces, Braid Movies and Beyond', *Knots and Quantum Gravity* ed. J. Baez, Clarendon Press, Oxford (1994) p.191–229.

[CF] L. Crane, I.B. Frenkel, 'Four-dimensional topological quantum field theory, Hopf categories, and the canonical bases', *J. Math. Phys.* **35** (1994) p.5136–5154.

[CKY] L. Crane, L.H. Kauffman, D. Yetter, 'State-Sum Invariants of 4-Manifolds, I', *preprint hep-th/9409167* (1994).

[CY] L. Crane, D. Yetter, 'A categorical construction of 4D topological quantum field theories', *Quantum Topology* World Scientific (1993) p.120–130.

[Fi] J.E. Fischer, '2–Categories and 2–Knots', *Duke Math. J.* (1994) **75** p.493–526.

[F] D.S. Freed, 'Higher algebraic structures and quantization', *Commun. Math. Phys.* **159** (1994) p.343–398.

[GK] S. Gelfand, D. Kazhdan, 'Invariants of three-dimensional manifolds', *preprint* (1994).

[GPS] R. Gordon, A.J. Power, R. Street, 'Coherence for tricategories', *Mem. Amer. Math. Soc., to appear.*

[HOMFLY] P. Freyd, D. Yetter, J. Hoste, W.B.R. Lickorish, K.C. Millett, A. Ocneanu, 'A new polynomial invariant of knots and links', *Bull. A.M.S.* **12** (1985) p.239–246.

[J1] V.F.R. Jones, 'A new knot polynomial and von Neumann algebras', *Notices A.M.S.* **33** (1986) p.219–225.

[J2] V.F.R. Jones, 'Hecke algebra representations of braid groups and link polynomials', *Ann. of Math.* **126** (1987) p.335–388.

[KV1] M.M. Kapranov, V.A. Voevodsky, 'The free n-category generated by a cube, oriented matroids and higher Bruhat orders', *Funct. Anal. Appl.* **25** (1991) p.62–65.

[KV2] M.M. Kapranov, V.A. Voevodsky, '2-Categories and Zamolodchikov tetrahedra equations', *Proc. Symp. Pure Math.* **56(2)** (1994) p.177–259.

[KT] V.M. Kharlamov, V.G. Turaev, 'On the definition of 2-category of 2-knots', *Preprint* (1994).

[K] G. Kelly, 'Coherence theorems for lax algebras and distributive laws', *Lect. Notes in Math.* **420** (1974) p.281–375.

[Ku] G. Kuperberg, 'Non-involutory Hopf algebras and 3–manifold invariants', *preprint* (1994).

[Lap] M. Laplaza, 'Coherence for distributivity', *Lect. Notes in Math.* **281** (1972) p.29–65.

[L1] R.J. Lawrence, 'On algebras and triangle relations', *Topological and geometrical methods in field theory* World Scientific (1992) p.429–447.

[L2] R.J. Lawrence, 'Triangulations, categories and extended topological field theories', *Quantum Topology* World Scientific (1993) p.191–208.

[L3] R.J. Lawrence, 'Yang-Baxter type equations and posets of maximal chains', *IHES preprint* **IHES/M/94/59** (1994).

[L4] R.J. Lawrence, 'Algebras and triangle relations', *J. Pure Appl. Alg.* **100** (1995).

[M] S. MacLane, 'Categories for the working mathematician', *GTM* **5** Springer-Verlag (1971).

[MS] Yu.I. Manin, V.V. Schechtman, 'Arrangements of hyperplanes, higher braid groups and higher Bruhat orders', *Adv. Studies in Pure Math.* **17** (1989) p.289–308.

[Ma] S. Matveev, 'Special spines of piecewise linear manifolds', *Mat. Sbornik* **92** (1973) p.282–293.

[P] U. Pachner, 'PL Homeomorphic manifolds are equivalent by elementary shellings', *Europ. J. Comb.* **12** (1991) p.129–145.

[PT] J.H. Przytycki, P. Traczyk, *preprint* (1985).

[Q] F. Quinn, 'Lectures on axiomatic topological field theory', *preprint* (1992).

[R] J. Roberts, 'Skein theory and Turaev-Viro invariants', *preprint* (1993).

[RT] N.Yu. Reshetikhin, V.G. Turaev, 'Invariants of 3-manifolds via link polynomials and quantum groups', *Invent. Math.* **103** (1991) p.547–597.

[Sch1] A.S. SCHWARZ, 'The partition function of degenerate quadratic functionals and Ray-Singer invariants', *Lett. Math. Phys.* **2** (1978) p.247–252.

[Sch2] A.S. SCHWARZ, 'New topological invariants arising in the theory of quantized fields', *Baku International Topological Conference, Abstracts (Part 2) Baku* (1987).

[S1] G.B. SEGAL, 'The definition of conformal field theory', *preprint.*

[S2] G.B. SEGAL, 'Two-dimensional conformal field theories and modular functors', *IXth International Congress on Mathematical Physics* Adam Hilger, Bristol (1989) p.22–37.

[St] R. STREET, 'The algebra of oriented simplices', *J. Pure Appl. Alg.* **49** (1987) p.283–335.

[T] V.G. TURAEV, 'Quantum Invariants of Knots and 3-manifolds', *Studies in Mathematics* **18** Walter de Gruyter, Berlin (1994).

[TV] V.G. TURAEV, O.YA. VIRO, 'State Sum Invariants of 3-manifolds and quantum 6j-symbols', *Topology* **31** (1992) p.865–902.

[W] E. WITTEN, 'Quantum Field Theory and the Jones Polynomial', *Commun. Math. Phys.* **121** (1989) p.351–399.

[Y] D. YETTER, 'State-sum invariants of 3-manifolds associated to Artinian semisimple tortile categories', *Topology Appl.* **58** (1994) p.47–80.

[Z1] A.B. ZAMOLODCHIKOV, 'Tetrahedra equations and integrable systems in three-dimensional space', *Sov. JETP* **52** (1980) p.325–336.

[Z2] A.B. ZAMOLODCHIKOV, 'Tetrahedron equations and the relativistic S-matrix of straight strings in $2+1$-dimensions', *Commun. Math. Phys.* **79** (1981) p.489–505.

DEPARTMENT OF MATHEMATICS, UNIVERSITY OF MICHIGAN, ANN ARBOR, MI 48109

Current Address: Institut des Hautes Études Scientifiques, F-91440 Bures-sur-Yvette, FRANCE.

E-mail: lawrence@math.lsa.umich.edu

Proceedings of Symposia in Applied Mathematics
Volume **51**, 1996

Vassiliev and Quantum Invariants of Braids

Dror Bar-Natan

ABSTRACT. We prove that braid invariants coming from quantum $gl(N)$ separate braids, by recalling that these invariants (properly decomposed) are all Vassiliev invariants, showing that all Vassiliev invariants of braids arise in this way, and reproving that Vassiliev invariants separate braids. We discuss some corollaries of this result and of our method of proof.

CONTENTS

1. Introduction

1.1. The result. Recall [**28, 29**] that given a list $R_1, \dots, R_n$ of representations of the lie algebra $gl(N)$ (or in fact, any other semi-simple Lie algebra) one can construct an n-component tangle invariant (the *Reshetikhin-Turaev invariant*), and in particular, an invariant

1991 *Mathematics Subject Classification.* Primary 57M25, 17B37; Secondary 05C99.
This work was supported by NSF grant DMS-92-03382.

This paper is available via anonymous file transfer from `ftp.math.harvard.edu`, user name `ftp`, subdirectory `dror`. Read the file `README` first. For easier access, point your WWW browser at `http://www.math.harvard.edu/HTML/Individuals/Dror_Bar-Natan.html`.

$J_{R_1,\dots,R_n}$ of n-strand pure braids, with values in $\mathrm{End}(R_1 \otimes \cdots \otimes R_n)[[\hbar]]$, the ring of formal power series[1] in the variable $\hbar$ with coefficients in $\mathrm{End}(R_1 \otimes \cdots \otimes R_n)$. The main goal of this paper is to prove the following theorem

THEOREM 1. *These invariants, coming from $gl(N)$ and all of its representations, separate pure braids.*

See remark 4.2 for a comment about (not necessarily pure) braids.

The main tools we will use in proving theorem 1 are Vassiliev invariants, chord diagrams, and weight systems. So before sketching the proof of theorem 1 in section 1.3, let us briefly recall these important notions.

1.2. Vassiliev invariants. Although originally defined only for knots, the notion of 'a Vassiliev invariant' ([**12, 13, 30, 31**]; see also [**2, 6, 8, 19**]) can be easily generalized to many other classes of 'knot-like' objects, such as braids, pure braids, tangles, links, string links, knotted graphs, knots in a 3-manifold, etc. The idea is always the same. Let $\mathcal{K}$ be a class of knot-like objects — a class of embeddings of oriented 1-dimensional objects in some 3-dimensional oriented space, perhaps satisfying some boundary conditions, considered modulo some reasonable notion of 'isotopy'. Let $V : \mathcal{K} \to \mathbf{A}$ be an invariant (under 'isotopy') with values in some Abelian group $\mathbf{A}$. It is always possible to extend V to 'knot-like objects with self-intersections' ('singular knot-like objects') by the formula

(1)

(We say that the double point is 'resolved into an overcrossing minus an undercrossing'). V is called *a Vassiliev invariant of type m* if its natural extension vanishes whenever it is evaluated on an object with more than m self-intersections:

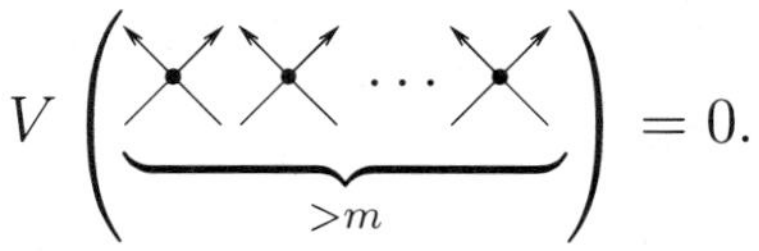

The relevance of Vassiliev invariants to our issue stems from the following two facts:

FACT 1. *(Lin [**22**], see also [**2, 6**]) The coefficient $J_{R_1,\dots,R_n,m}$ of $\hbar^m$ in $J_{R_1,\dots,R_n}$ is a Vassiliev invariant of type m.*

[1] The Laurent polynomials in the formal parameter q of [**28, 29**] become formal power series in $\hbar$ upon the substitution $q = e^{\hbar}$.

FACT 2. *(Bar-Natan* [**3**]*, see also Kohno* [**18**]*) Vassiliev invariants of (pure) braids separate (pure) braids.*

For the convenience of the reader, we will sketch a proof of fact 2 in section 3.

Differences are in many ways similar to derivatives, and as V  is defined to be a difference, one can think of Vassiliev invariant of type m as invariants whose higher-than-m derivatives vanish, or, as 'polynomials of degree at most m'. Just as in the case of degree m polynomials, the mth order 'derivatives' of a type m Vassiliev invariant are 'constant'. More precisely, if our knot-like object K has precisely m self-intersections, then by (1) and the definition of Vassiliev invariants, one can replace overcrossings in K by undercrossings (and vice versa) freely, without changing the values of any type m Vassiliev invariant V. This means (at least in the case where the ambient space is simply connected) that V doesn't really depend on the topology of K, but rather it depends only on the combinatorial object defined by the parameter space S of K together with the pairs of points in S that map into each of the self-intersections of K. Such pairs of points on S are usually signified by drawing a 'chord' connecting them, and the resulting combinatorial object is called *a chord diagram.*

The simplest and best known (see e.g. [**2, 6, 8, 19**]) example is that of oriented knots in oriented space. In that case, the parameter space is an oriented circle (conventionally oriented counterclockwise when drawn), and an example for a chord diagrams is in figure 1.

Figure 1. A chord diagram of degree 4.

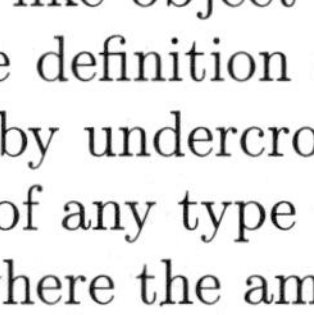

The example that will be of interest for us is that of $\mathcal{K} = \{n\text{-strand}$ pure braids$\}$. In that case, (degree m) chord diagrams are diagrams made of n vertical directed lines ('strands') and m horizontal lines ('chords') connecting them, as in figure 2.

Thus, to every **A**-valued Vassiliev invariant V of type m of pure braids, corresponds a map $W_m(V) : \mathcal{G}_m\mathcal{D}^{pb} \to \mathbf{A}$ defined on the **Z**-module $\mathcal{G}_m\mathcal{D}^{pb}$ freely generated by all degree m (pure braid) chord diagrams. Notice that by stacking chord diagrams vertically, $\mathcal{D}^{pb} \stackrel{\text{def}}{=} \bigoplus_m \mathcal{G}_m\mathcal{D}^{pb}$ becomes a (non-commutative) graded algebra. As an algebra $\mathcal{D}^{pb}$ is generated by the degree 1 chord diagrams t^{ij}, $(1 \leq i \neq j \leq n)$, given by

Figure 2. A singular 3-strand pure braid with 4 self-intersections, and the corresponding chord diagram.

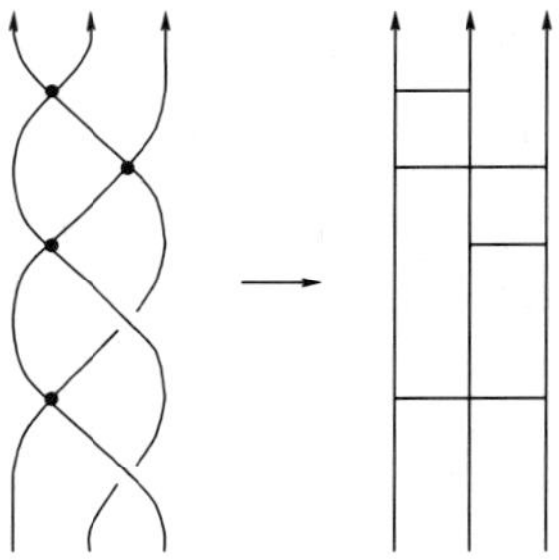

$$t^{ij} = t^{ji} =$$ 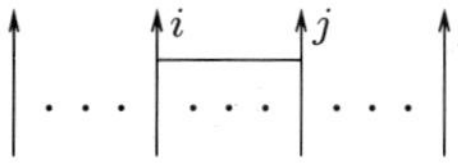

FACT 3. *$W_m(V)$ vanishes on the degree m piece of the double-sided ideal $\mathcal{I}$ of $\mathcal{D}^{pb}$ generated by the relations*

$$t^{ij}t^{kl} = t^{kl}t^{ij} \quad \textit{when} \quad |\{i,j,k,l\}| = 4, \tag{2}$$

$$\left[t^{ik} + t^{jk}, t^{ij}\right] = 0 \quad \textit{when} \quad |\{i,j,k\}| = 3. \tag{3}$$

Indeed, relation 2 just says that double points 'can be moved across each other', as in figure 3, while relation 3 is just the $4T$ relation of [**2**], in a slightly disguised form. Its proof is sketched in figure 4.

Figure 3. The two singular braids (which may be parts of 'bigger' pure braids) displayed here are equivalent, implying relation 2.

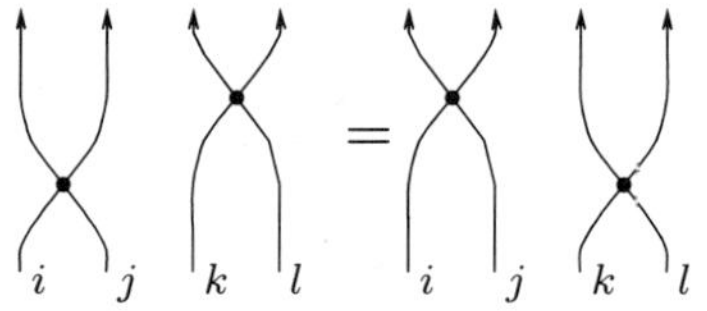

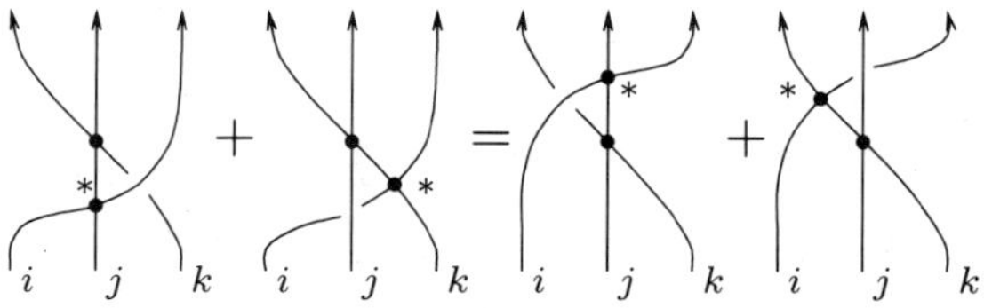

Figure 4. The chord diagrams corresponding to the four singular braids displayed here are the four terms of (3). When the double points marked by a $*$ is resolved into overcrossings and undercrossings as in 1, the resulting 8 singular braids cancel in pairs.

We set $\mathcal{A}^{pb} = \mathcal{D}^{pb}/\mathcal{I}$, and then fact 3 just says that to every **A**-valued Vassiliev invariant V of pure braids corresponds a map (a *weight* *system)* $W_m(V) : \mathcal{G}_m\mathcal{A}^{pb} \to \mathbf{A}$. Over the rationals, the converse is also true:

FACT 4. *Every weight system comes from an invariant. More precisely, if* $\mathbf{A}$ *is a* $\mathbf{Q}$*-module and* $W : \mathcal{G}_m\mathcal{A}^{pb} \to \mathbf{A}$ *is arbitrary, then there exists a* $\mathbf{A}$*-valued Vassiliev invariant* V *of pure braids for which* $W = W_m(V)$*. The invariant* V *is determined uniquely up to invariants of lower type.*

Over the real numbers, fact 4 can be proven using the 'Kontsevich integral formula' of [**19**] (see also [**2**]). Over the rationals it can be deduced from Drinfel'd's work [**10, 11**] on quasi-Hopf algebras. See e.g. [**1, 4, 9, 15, 21, 23, 27**].

1.3. Sketch of the proof. Fact 2 (see section 3) implies that in order to prove theorem 1, it is enough to prove that all Vassiliev invariants of pure braids come (as in fact 1) from the $gl(N)$ invariants $J_{R_1,\dots,R_n,m}$. Facts 3 and 4 imply that it is enough to do that on the level of weight systems; that is, it is enough to prove that the weight systems corresponding to (various traces of) the $J_{R_1,\dots,R_n,m}$'s span the space $\mathcal{W}^{pb} = (\mathcal{A}^{pb})^*$ of all weight systems. This is exactly corollary 2.6 of section 2.5. In section 4 we will prove some corollaries of theorem 1 and discuss some related questions.

REMARK 1.1. The technique used in this paper appears to be a part of a pattern — statements about knot polynomial or quantum group invariants of knots become simpler when restated in terms of Vassiliev invariants, weight systems, and chord diagrams. Chord diagrams (which are just abstract graphs) are much more manageable objects than knots (whose embedding into space matters), and so complicated facts about knots become provable once stated in this simpler language. Perhaps an even better example for this principle is the proof of the Melvin-Morton-Rozansky conjecture in [**5**].

REMARK 1.2. The two main problems in the theory of Vassiliev invariants are:

- Do Vassiliev invariants separate knotted objects (in some class $\mathcal{K}$)?
- And do they all come from lie algebras?

In the case of $\mathcal{K} = \{\text{knots}\}$, we know (see e.g. [**20**]) that the answer cannot be "yes" for both questions, though we don't know any of the answers. Fact 2 says that the answer to the first question is "yes" in the case of $\mathcal{K} = \{\text{pure braids}\}$, and the main point in section 2 is to show that the answer to the second question is also "yes" in that case.

Another case in which both questions have an affirmative answer is $\mathcal{K} = \{\text{homotopy string links}\}$. See [**3**].

1.4. Acknowledgement. I would like to thank S. Garoufalidis, V. Jones, D. Kazhdan, D. Long, P. M. Melvin, H. R. Morton, and N. Reshetikhin for their many useful comments, and Yael for babysitting Assaf over the weekend in which most of this article was written.

2. All Vassiliev invariants of pure braids come from $gl(N)$

In this section we will prove (as promised in section 1.3) that the weight systems corresponding to the $gl(N)$ invariants $J_{R_1,\dots,R_n,m}$ span the space of all pure braid weight systems. Denote the weight system of $J_{R_1,\dots,R_n,m}$ by $W_{R_1,\dots,R_n}$ (suppressing the degree m, which anyway can be read from the degree of the chord diagrams being fed into $W_{R_1,\dots,R_n}$). Our first step is to better understand the $\mathrm{End}(R_1 \otimes \cdots \otimes R_n)$-valued weight system $W_{R_1,\dots,R_n}$, for some fixed list $R_1, \dots, R_n$ of representations of $gl(N)$.

2.1. The weight system $W_{R_1,\dots,R_n}$. Let R be the defining (N-dimensional) representation of $gl(N)$, and let $\{M_a\}_{a=1}^{N^2}$ be a basis of $gl(N)$, orthonormal with respect to the invariant metric $\langle M_a, M_b\rangle = \mathrm{tr}_R(M_a M_b)$. Define a map $\mathcal{D}^{pb} \to \mathrm{End}(R_1 \otimes \cdots \otimes R_n)$ by extending the map

$t^{ij} =$ 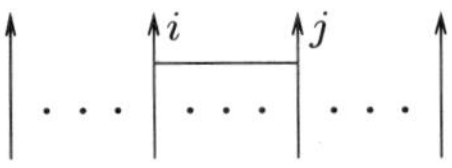

$$\mapsto \sum_{a=1}^{N^2} 1 \otimes \cdots \otimes 1 \otimes \underset{i}{M_a} \otimes 1 \cdots 1 \otimes \underset{j}{M_a} \otimes 1 \otimes \cdots \otimes 1 \quad (4)$$

to be an algebra morphism.

FACT 5. 1. *(Kohno* [**17**], *see also* [**2**, proposition 2.11]*) The above defined map $\mathcal{D}^{pb} \to \mathrm{End}(\otimes_i R_i)$ descends to a well defined map $\mathcal{A}^{pb} \to \mathrm{End}(\otimes_i R_i)$.*
2. *(Piunikhin* [**26**], *see also* [**2**, remark 4.8]*) The resulting map $\mathcal{A}^{pb} \to \mathrm{End}(\otimes_i R_i)$ is the weight system $W_{R_1,\dots,R_n}$.*

2.2. The defining representation of $gl(N)$. Let us now specialize to the case when $R_{1,\dots,n} = R$, the defining (N-dimensional) representation of $gl(N)$.

FACT 6. *(See e.g.* [**2**, exercise 6.36]*) The tensor $t = \sum_a M_a \otimes M_a \in \mathrm{End}(R \otimes R)$ is the 'crossed identity' tensor given by the formula*

$$t(v \otimes w) = w \otimes v.$$

It is consistent with the notation in (4) to denote the identity operator by a directed line, and then fact 6 becomes the statement

$t =$ 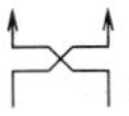 ,

and (4) becomes

$W_{R,\dots,R}:$ 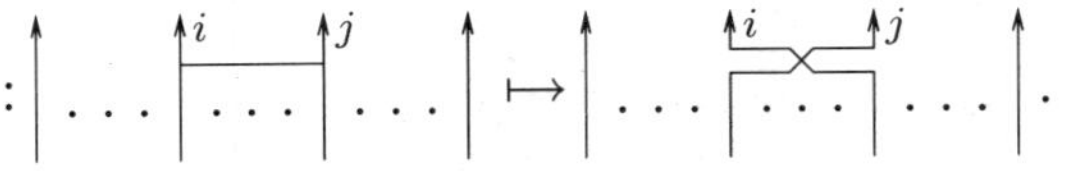

We can get numerical valued weight systems out of $W_{R,\dots,R}$ by composing it with various traces $\mathrm{End}(R^{\otimes n}) \to \mathbf{C}$. I.e., if $\sigma \in \mathcal{S}_n$ is a permutation of $1, \dots, n$, it defines in a natural way a permutation operator (also denoted by the letter σ) in $\mathrm{End}(R^{\otimes n})$, and then one can define $\mathrm{tr}_\sigma : \mathrm{End}(R^{\otimes n}) \to \mathbf{C}$ by $\mathrm{tr}_\sigma E = \mathrm{tr}\, \sigma \circ E$. Let W_σ denote the numerical valued weight system $\mathrm{tr}_\sigma \circ W_{R,\dots,R}$.

Proposition 2.1. *Let $D \in \mathcal{D}^{pb}$ be a chord diagram. Then $W_\sigma(D)$ can be computed in three steps as below (and as in figure 5):*

1. *Append the permutation σ to the top of D, and 'close around'.*
2. *Replace all chords in D by 'crossings'.*
3. *Count the number c of components in the resulting diagram. $W_\sigma(D)$ is then N^c.*

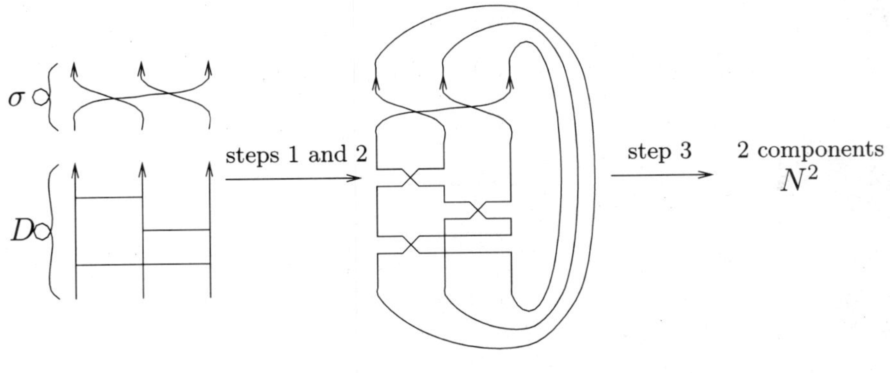

Figure 5. The three steps of computing $W_\sigma(D)$.

2.3. Tensor powers of the defining representation. The following definition and fact allow us to compute $W_{R_1,\dots,R_n}(D)$ whenever D is a chord diagram and $\{R_i = R^{\otimes k_i}\}$ are tensor powers of the defining representation.

DEFINITION 2.2. Let $k = (k_1, \dots, k_n)$ be a sequence of non negative integers, let $|k| = \sum_i k_i$ be their sum, and let $l_i = \sum_{\alpha<i} k_\alpha$. Define an algebra morphism $\Delta^k : \mathcal{D}^{pb}_n \to \mathcal{D}^{pb}_{|k|}$ by setting

$$\Delta^k(t^{ij}) = \sum_{i'=l_i+1}^{l_{i+1}} \sum_{j'=l_j+1}^{l_{j+1}} t^{i'j'}.$$

This map descends to a morphism (called by the same name) $\Delta^k : \mathcal{A}^{pb}_n \to \mathcal{A}^{pb}_{|k|}$.

Δ^k can be thought of as replacing the ith strand in D by a bundle of k_i strands (for each i) and summing over all possible ways of 'lifting' the chords from their original strand to the bundle that replaced it. For example, $\Delta^{(22)}(t^{12}) = t^{13} + t^{14} + t^{23} + t^{24}$, or

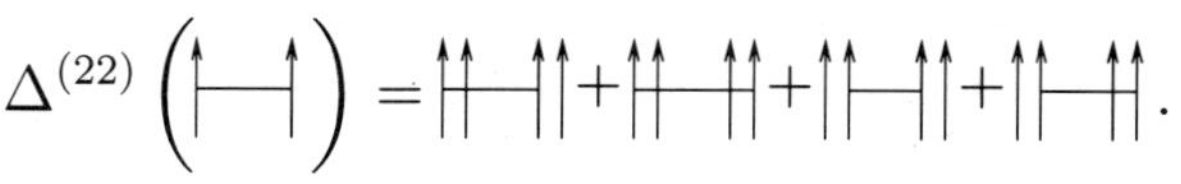

FACT 7. *(See proposition 6.18 of* [**2**]*) The following diagram commutes:*

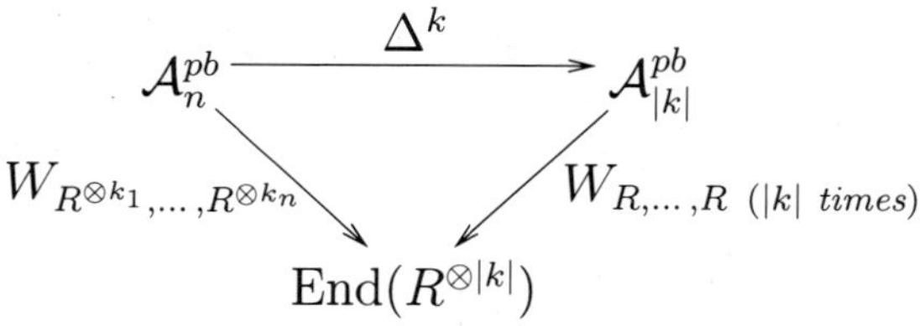

Given a list $k = (k_1, \dots, k_n)$ and a permutation $\sigma \in \mathcal{S}_{|k|}$ of the integers 1 through $|k|$, we consider the numerical valued weight system

$$W_{k,\sigma} = \mathrm{tr}_\sigma \circ W_{R^{\otimes k_1}, \dots, R^{\otimes k_n}}.$$

Fact 7 and proposition 2.1 show that we can compute $W_{k,\sigma}(D)$ for any chord diagram D following the steps below:

1. For each i, replace the ith strand in D by a bundle of k_i strands and consider all possible ways $\{D_\alpha\}$ of lifting all chords.
2. For each α, append the permutation σ at the top of D_α, close around, replace all chords by crossings, and call the result D'_α.
3. $W_{k,\sigma}(D)$ is $\sum_\alpha N^{c_\alpha}$, where c_α is the number of components in D'_α.

2.4. Paths. It is easy to see that $W_{k,\sigma}(D)$ does not change if $\sigma \in \mathcal{S}_{|k|}$ is conjugated by a permutation in $\mathcal{S}_{k_1} \times \cdots \times \mathcal{S}_{k_n} \subset \mathcal{S}_{|k|}$, as this just corresponds to permuting the strands within each bundle. A *path*, defined below, is basically a pair (k, σ), with the redundancy in σ removed.

DEFINITION 2.3. A connected path is a word in the n letters S_1 through S_n. A path is an unordered list (possibly with multiplicities) of connected paths.

Figure 2.4 explains by an example how a pair (k, σ) as above determines a path P (up to cyclically permuting the letters of each word in P), and how a path P determines a pair (k, σ) (up to conjugating σ by a permutation in $\mathcal{S}_{k_1} \times \cdots \times \mathcal{S}_{k_n}$).

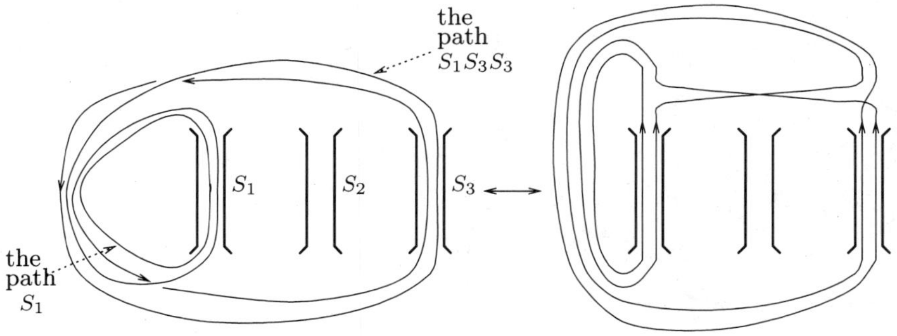

Figure 6. The path $P = (S_1, S_1S_3S_3)$ and the pair $(k = (2,0,2),\ \sigma = (1)(234))$ correspond.

We thus find that to every path P corresponds a weight system $W_P = W_{k,\sigma}$, where (k, σ) correspond to P as in figure 2.4. The algorithm in section 2.3 becomes the following algorithm for computing $W_P(D)$, where D is a chord diagram of degree m:

1. For each connected component of P write a long interval, subdivided into shorter subintervals corresponding to the letters making up that component. Mark sites corresponding to the integers $1, \ldots, m$ in order along each of the subintervals. For example, if $m = 2$, the path $P = (S_1, S_1S_3S_3)$ becomes:

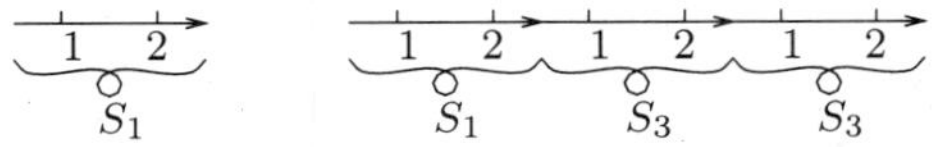

(5)

2. Consider all liftings $\{D_\alpha\}$ of D to the picture just drawn, where each end of each chord is lifted to one of the sites in the picture, so that if the (say) 7th chord in D is t^{23}, then its ends are lifted to sites on marked by the integer 7 on subintervals corresponding to the letters S_2 and S_3. For example, the chord diagram $D = t^{13}t^{13}$ has 16 possible liftings to the path $(S_1, S_1S_3S_3)$, as the first t^{13} in D can be lifted to become either one of the 4 chords in the first

figure below, and the second t^{13} in D can be lifted to become either one of the 4 chords in the second figure:

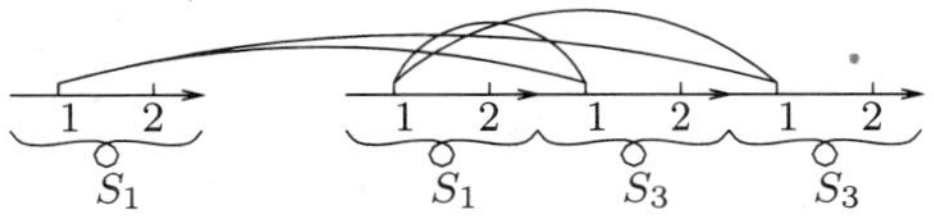

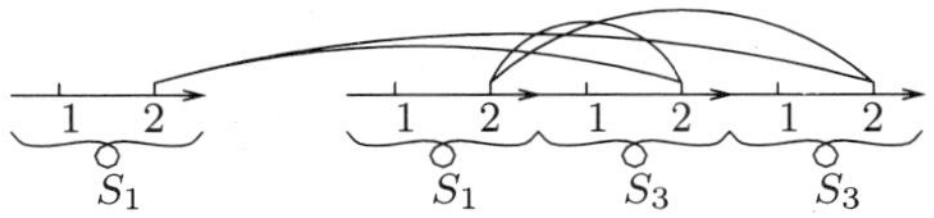

As a second example, notice that the chord diagram $t^{13}t^{23}$ has no liftings to the path $(S_1, S_1S_3S_3)$, as that path does not pass through the strand S_2 at all, and there's nowhere to lift t^{23} to.

3. Replace all chords in each D_α by 'bridges' (notice that the 'crossings' of section 2.3 become 'bridges' in the current picture), erase all markings, close all intervals into loops, and call the result D'_α:

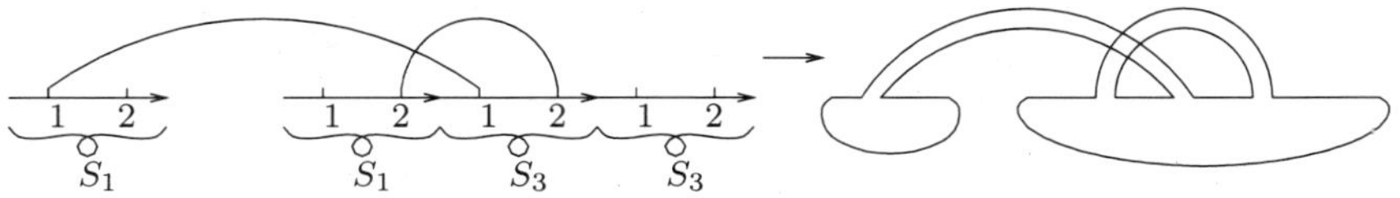

4. $W_P(D)$ is $\sum_\alpha N^{c_\alpha}$, where c_α is the number of components in D'_α.

2.5. The conclusion of the proof. Define the *order* of a chord t^{ij} to be $\max(i,j)$. We say that a chord diagram is *non-decreasing* if its chords appear in non-decreasing order. We say that it is *flat* if all the chords in it are of the same order. For example, $t^{13}t^{23}t^{13}$ is flat and non-decreasing, $t^{12}t^{23}t^{13}$ is non-decreasing but not flat, and $t^{13}t^{12}$ is neither flat nor non-decreasing. The following fact is a trivial consequence of the relations (3) and (2) defining $\mathcal{A}^{pb}$.

FACT 8. *(Drinfel'd* **[11]***) $\mathcal{A}^{pb}$ is generated by non-decreasing chord diagrams.*

The following two propositions make the key technical observation of this paper.

PROPOSITION 2.4. *The weight systems W_P separate flat chord diagrams. More precisely, if $D_\nu(i_1,\dots,i_m) = t^{i_1\nu}\cdots t^{i_m\nu}$ is a flat chord diagram of degree m (that is, $i_\alpha < \nu$ for all α) and $P_\nu(j_1,\dots,j_m)$ (with $j_\alpha < \nu$ for all α) is the connected path $S_{j_m}S_{j_{m-1}}\cdots S_{j_1}S_\nu$, and if $C_{N,p}$ is the operator that maps a polynomial in N to the coefficient of N^p in it, then*

$$C_{N,m+1}W_{P_\nu(j_1,\dots,j_m)}(D_\nu(i_1,\dots,i_m)) = \begin{cases} 1 & \text{if } i_\alpha = j_\alpha \text{ for all } \alpha, \\ 0 & \text{otherwise.} \end{cases}$$

PROOF. Consider the diagrams D_α and the corresponding numbers c_α that appear when $W_{P_\nu(j_1,\dots,j_m)}(D_\nu(i_1,\dots,i_m))$ is computed using the algorithm of section 2.4. It is easy to show that the maximal possible value for the c_α's is $m+1$, and that this maximum is attained iff the corresponding diagram D_α has no chord intersections:

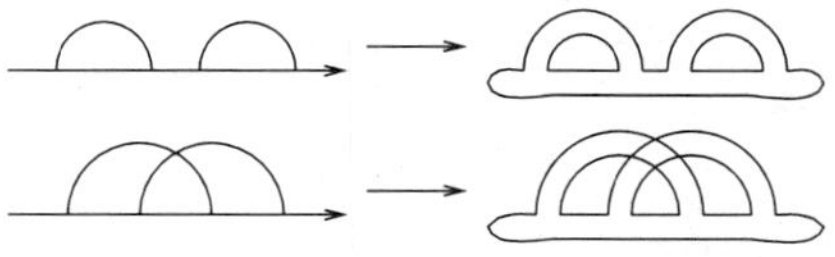

Therefore, we only care about those D_α's in which there are no chord intersections. Notice that the right ends of the chords in $D_\nu(i_1,\dots,i_m)$ can only be lifted (in order) to the last interval (S_ν) in $P_\nu(j_1,\dots,j_m)$ — they simply have nowhere else to go. The left ends of these chords should be lifted to one of the earlier intervals in $P_\nu(j_1,\dots,j_m)$, and so far the picture is:

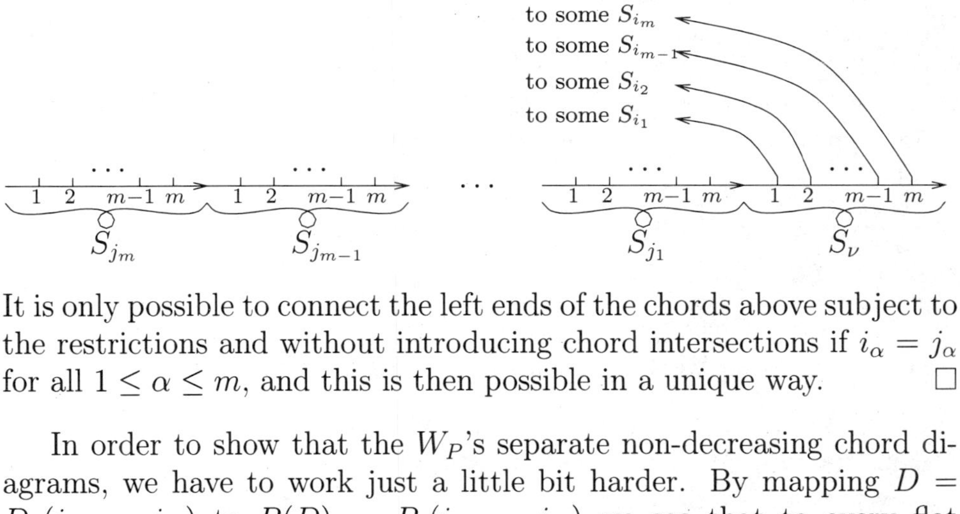

It is only possible to connect the left ends of the chords above subject to the restrictions and without introducing chord intersections if $i_\alpha = j_\alpha$ for all $1 \le \alpha \le m$, and this is then possible in a unique way. $\square$

In order to show that the W_P's separate non-decreasing chord diagrams, we have to work just a little bit harder. By mapping $D = D_\nu(i_1,\dots,i_m)$ to $P(D) = P_\nu(i_1,\dots,i_m)$ we see that to every flat chord diagram D corresponds a connected path $P(D)$. This map can be extended to non-decreasing chord diagrams — simply write any non-decreasing chord diagram D as an increasing product of flat chord diagrams, and let $P(D)$ be the list of connected paths corresponding to the flat parts of D (including a connected path S_ν for each $1 \le \nu \le n$ for which the order ν part of D is empty). Let the profile of a diagram D be the sequence $(l_n, l_{n-1}, \dots, l_1)$, where l_ν is the number of chords in D which are of order ν. Let us order all degree m chord diagrams using some order $\prec$ that refines the lexicographic order on their profiles. The following proposition is a generalization of proposition 2.4, and its proof is very similar.

PROPOSITION 2.5. *The W_P's separate non-decreasing chord diagrams. More precisely, if D_1 and D_2 are chord diagrams of degree m, then*

$$C_{N,m+n}W_{P(D_1)}(D_2) = \begin{cases} 1 & \text{if } D_1 = D_2, \\ 0 & \text{if } D_1 \prec D_2. \end{cases}$$

PROOF. Consider the diagrams D_α and the numbers c_α that appear when $W_{P(D_1)}(D_2)$ is computed using the algorithm of section 2.4. It is easy to show that the maximal possible value for the c_α's is $m+n$, and that this maximum is attained iff the corresponding diagram D_α has no chord intersections, and no chords connecting different components of $P(D_1)$. Therefore, we only care about those D_α's in which there are no chord intersections and no such connecting chords. If the profile of D_1 is lexicographically smaller than the profile of D_2, there is simply no room to fit a lifting of D_2 on $P(D_1)$ without chord intersections or connecting chords. If the profiles of D_1 and D_2 are the same, the flat components of D_2 must be lifted to the corresponding connected paths in $P(D_1)$, and the same argument as in the proof of proposition 2.4 shows that such a lifting (having no chord intersections or connecting chords) exists iff $D_1 = D_2$, and in that case it is unique. □

COROLLARY 2.6. *The weight systems corresponding to (traces of) the $gl(N)$ invariants $J_{R_1,\dots,R_n,m}$ span the space $\mathcal{W}^{pb} = (\mathcal{A}^{pb})^*$ of all weight systems.* □

3. Vassiliev invariants separate pure braids

Just for the amusement of the reader, we include here the 'moral reason' for why Vassiliev invariants separate pure braids. As proofs of this fact are available elsewhere (see e.g. [**3, 18**]), we will leave the details of the proof presented here as an exercise to the reader.

FACT 9. 1. *Every pure braid has a unique presentation as a 'combed braid', as in figure 7. Equivalently, the group K_n of pure braids on n strands is a semi-direct product of free groups:*

$$K_n = F_1 \ltimes F_2 \ltimes \cdots \ltimes F_{n-1}. \tag{6}$$

2. *(Kohno [**17**]) The non-decreasing chord diagrams form a basis of $\mathcal{A}^{pb}$. Equivalently, the algebra $\mathcal{A}_n^{pb}$ of chord diagrams on n strands is a semi-direct product of free associative algebras:*

$$\mathcal{A}_n^{pb} = FA_1 \ltimes FA_2 \ltimes \cdots \ltimes FA_{n-1}. \tag{7}$$

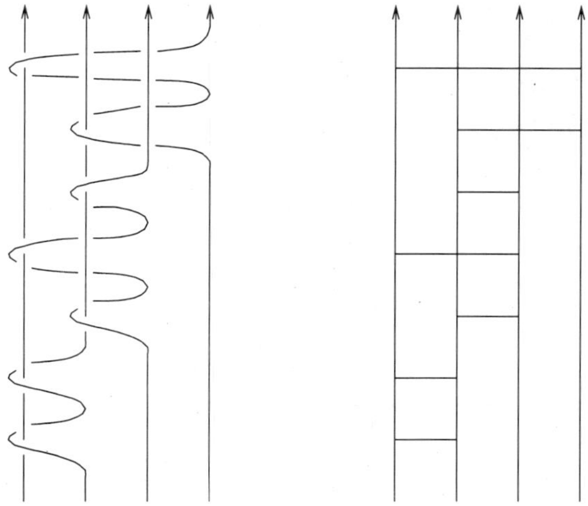

Figure 7. Left: In a combed braid, the first strand is always straight. The second twists around the first some number of times, and then becomes straight. The third waits patiently until the second finishes twisting around the first, tangles a bit between the first and the second, and then becomes straight. The fourth waits for the third to be done, tangles, etc. Right: a non-decreasing chord diagram.

PROOF. 1. Every braid can be 'combed' by induction. Assume the first four strands are already combed as in figure 7, and that a fifth strand tangles between them. Think of the first four strands as made of copper wires and grease them very well. Think of the fifth strand as made of soft spaghetti, and use a fan to blow strong wind from the bottom up. The spaghetti wants to fly up, and the copper wires are very smooth (and sloped upwards) so they can't stop it from doing so. When all the spaghetti (except the very beginning, which is tied to the bottom plane) reaches the top of the figure, freeze it in place. It is now a path in the the plane minus four points (the four other strands). Present it as a product of generators, replace it by a copper wire, and you are ready to deal with the sixth strand. For a formal proof, see e.g. [**7**].

2. Immediate from fact 8 (that non-decreasing chord diagrams generate $\mathcal{A}^{pb}$) and from the main technical observation of this paper, proposition 2.5.

□

EXERCISE 3.1. The (formal) Knizhnik-Zamolodchikov connection [**16, 17, 19, 2**] defines a 'holonomy map' $Z : K_n \to \mathcal{A}_n^{pb}$. Use simple properties of Z (or of any of the other 'universal Vassiliev invariants' constructed in [**1, 4, 9, 15, 21, 27**]) and the obvious similarity between equation 6 and equation 7 to show that Z is injective. Composing with linear functionals on $\mathcal{A}_n^{pb}$, one gets exactly the Vassiliev invariants.

4. Corollaries

4.1. The HOMFLY polynomial and braids. It is well known (see e.g. [**28, 29**] that $gl(N)$ in its defining representation corresponds to the HOMFLY polynomial (in some parametrization), and that the higher representations of $gl(N)$ correspond to various cabling operations applied to the HOMFLY polynomial (for a similar situation, see [**25**]).

COROLLARY 4.1. *A pure braid is determined by the HOMFLY polynomials of all closures into links of all of its cablings. (We allow an arbitrary permutation of the strands before closing).*

REMARK 4.2. The above corollary remains true even if the words 'pure braid' are replaced by the word 'braid'. Indeed, it is easy to read the number of components of a link from its HOMFLY polynomial, and knowing this number for all possible closures of a braid (using all possible permutations of the strands) determines the permutation σ underlying that braid, as that number is maximal only if the closure is done using the permutation σ^{-1}.

4.2. Braids and string links. The Reshetikhin-Turaev $gl(N)$ invariants $J_{R_1,\dots,R_n}$ were originally defined as tangle invariants, and, in particular, they extend to n-component string links (for a definition, check [**14**] or [**3**]). Clearly, the same holds for the various traces of the $J_{R_1,\dots,R_n}$'s, and as these span the space of all Vassiliev invariants of pure braids, it follows that

COROLLARY 4.3. *A Vassiliev invariant of pure braids can always be extended to become a Vassiliev invariant (of the same type) of string links.*

On the level of chord diagrams (which are dual to invariants), this corollary implies that chord diagrams of pure braids inject into chord diagrams of string links. Recall that the latter space, $\mathcal{A}^{sl}$, discussed in more detail in [**3**], is the space of all diagrams of the form

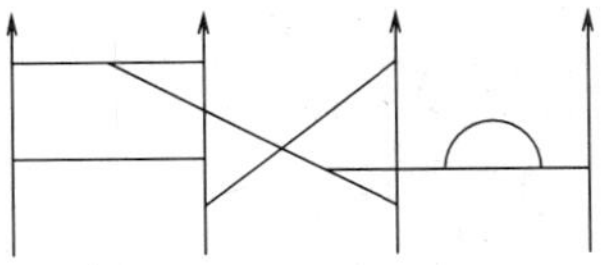

(non-horizontal chords and oriented internal trivalent vertices are allowed, apparent quadrivalent vertices in the planar projection are not true vertices), modulo the STU relation,

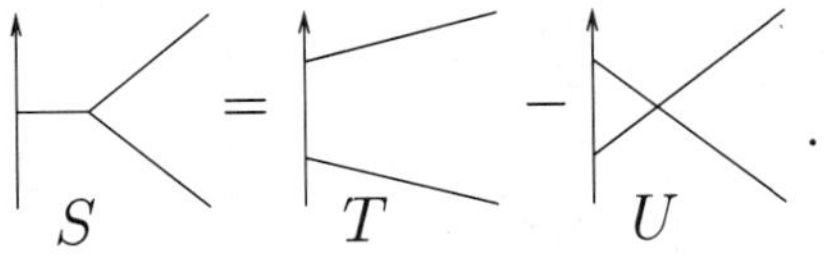

Corollary 4.4. *The obvious map $\mathcal{A}^{pb} \to \mathcal{A}^{sl}$ is an injection.*

In fact, curiosity whether corollary 4.4 holds is what lead me to the investigations described in this paper.

References

[1] D. Altschuler & L. Freidal, *On universal Vassiliev invariants*, February 1994, preprint.
[2] D. Bar-Natan, *On the Vassiliev knot invariants*, Topology **34** 423–472 (1995).
[3] ______, *Vassiliev homotopy string link invariants*, Jour. of Knot Theory and its Ramifications, to appear.
[4] ______, *Non-associative tangles*, to appear in the Georgia Inter. Topology Conf. proceedings.
[5] ______ & S. Garoufalidis, *On the Melvin-Morton-Rozansky conjecture*, July 1994, preprint.
[6] J. S. Birman, *New points of view in knot theory*, Bull. Amer. Math. Soc. **28** (1993) 253–287.
[7] J. S. Birman, *Braids, links, and mapping class groups*, Princeton Univ. Press, Princeton, 1975.
[8] ______ & X-S. Lin, *Knot polynomials and Vassiliev's invariants*, Inv. Math. **111** (1993) 225–270.
[9] P. Cartier, *Construction combinatoire des invariants de Vassiliev-Kontsevich des nœuds*, C. R. Acad. Sci. Paris **316** Série I (1993) 1205–1210.
[10] V. G. Drinfel'd, *Quasi-Hopf algebras*, Leningrad Math. J. **1** (1990) 1419–1457.
[11] ______, *On quasitriangular Quasi-Hopf algebras and a group closely connected with Gal*($\bar{\mathbb{Q}}/\mathbb{Q}$), Leningrad Math. J. **2** (1991) 829–860.
[12] M. Gusarov, *A new form of the Conway-Jones polynomial of oriented links*, in *Topology of manifolds and varieties* (O. Viro, editor), Amer. Math. Soc., Providence 1994, 167–172.
[13] ______, *On n-equivalence of knots and invariants of finite degree*, in *Topology of manifolds and varieties* (O. Viro, editor), Amer. Math. Soc., Providence 1994, 173–192.

[14] N. Habegger & X-S. Lin, *The Classification of links up to link-homotopy*, J. Amer. Math. Soc. **3** (1990) 389–419.

[15] C. Kassel, *Quantum groups*, Springer-Verlag GTM **155**, Heidelberg 1994.

[16] V. G. Knizhnik and A. B. Zamolodchikov, *Current algebra and Wess-Zumino model in two dimensions*, Nucl. Phys. **B247** (1984) 83–103.

[17] T. Kohno, *Monodromy representations of braid groups and Yang-Baxter equations*, Ann. Inst. Fourier **37** (1987) 139–160.

[18] ———, *Vassiliev invariants and de-Rham complex on the space of knots*, Tokyo Univeristy preprint, 1994.

[19] M. Kontsevich, *Vassiliev's knot invariants*, Adv. in Sov. Math., **16(2)** (1993), 137–150.

[20] G. Kuperberg, *Detecting knot invertibility*, University of Chicago preprint, February 1995.

[21] T. Q. T. Le & J. Murakami, *The universal Vassiliev-Kontsevich invariant for framed oriented links*, Max-Planck-Institut Bonn preprint, 1993.

[22] X-S. lin, *Vertex models, quantum groups and Vassiliev's knot invariants*, Columbia Univ. preprint, 1991.

[23] ———, *Knot invariants and iterated integrals*, Columbia Univ. preprint, 1994.

[24] ———, *Braid algebras, trace modules and Vassiliev invariants*, Columbia Univ. preprint, 1994.

[25] H. R. Morton and P. Strickland, *Jones polynomial invariants of knots and satellites*, Math. Proc. Camb. Phil. Soc. **109** (1991) 83–103.

[26] S. Piunikhin, *Weights of Feynman diagrams, link polynomials and Vassiliev knot invariants*, Moscow State Univ. preprint, 1992.

[27] ———, *Combinatorial expression for universal Vassiliev link invariant*, to appear in Commun. Math. Physics.

[28] N. Yu. Reshetikhin, *Quasitriangle Hopf algebras and invariants of tangles*, Leningrad Math. J. **1** (1990) 491–513.

[29] ——— & V. G. Turaev, *Ribbon graphs and their invariants derived from quantum groups*, Commun. Math. Phys. **127** (1990) 1–26.

[30] V. A. Vassiliev, *Cohomology of knot spaces*, Theory of Singularities and its Applications (Providence) (V. I. Arnold, ed.), Amer. Math. Soc., Providence, 1990.

[31] ———, *Complements of discriminants of smooth maps: topology and applications*, Trans. of Math. Mono. **98**, Amer. Math. Soc., Providence, 1992.

Department of Mathematics, Harvard University, Cambridge, MA 02138

E-mail address: dror@math.harvard.edu

Proceedings of Symposia in Applied Mathematics
Volume **51**, 1996

THE MODERN LEGACIES OF THOMSON'S ATOMIC VORTEX THEORY IN CLASSICAL ELECTRODYNAMICS

SAMUEL J. LOMONACO, JR.

Dedicated to the memory of Ralph H. Fox

ABSTRACT. The physicist Sir William Thomson (also known as Lord Kelvin) proposed in 1867 [56] [57] [58] that physical atoms were knotted vortex tubes in the then postulated all pervasive fluid called ether. The physicist Peter Guthrie Tait became so enamored with Thomson's theory that he undertook a study of the mathematical properties of knots, thus giving birth to the field of knot theory.

Although scientific evidence has since shown conclusively that physical atoms are by no means knotted vortices in the sense of Thomson, Thomson's theory has fragmented and relatively recently reemerged in many much more sophisticated forms in both classical and nonclassical physics. With the work of Jones, Witten, and others, knot theory has now begun to reassociate on a serious basis with its long lost ancestor, physics [28].

The first section of this paper gives a brief survey of the early Thomson atomic vortex theory as it developed within the James Clerk Maxwell milieu. The paper then focuses on the modern legacies of this theory in classical electrodynamics. In particular, the second and third sections of this paper look at some of the recent developments on knotted magnetic vortices and knotted electrostatic vortices, respectively.

In summary, this paper focuses on the study of the electromechanical behavior of knotted tubes of electrical charge and magnetic flux. Surprisingly, even within classical physics (more specifically, classical electrodynamics), there are many important unresolved questions about such objects. Many of these questions are relevant to such diverse fields as plasma physics, polymer physics, molecular biology, and, of course, knot theory.

Date. April 23, 1995.

1991 *Mathematics Subject Classification.* Primary 57-02, 57M25, 78-02, 78A30, 76W05, 53C80; Secondary 01-02, 78-03, 45K05, 22E70.

Key words and phrases. Knots, Electrodynamics, Electromagnetism, Energy, Magnetic Knots, Electrostatic Knots, Minimum Energy Knots, Vortices, Gauge Theory.

Partially supported by the L-O-O-P Fund.

I would like to thank Mike Freedman and Keith Moffatt for their helpful comments.

1. **Introduction:** A TALE OF TWO CENTURIES BEGINS WITH KELVIN'S KNOTTED VORTICES

We begin our tale of two centuries in the late nineteenth century, a time when the atomic structure of matter was just beginning to unfold. Herman von Helmholtz had just published a paper [26] on what James Clerk Maxwell referred to as "water twists." In this paper, Helmholtz proved the surprising result that, within an incompressible, inviscid, constant density fluid, fluid vortices are actually permanent and indivisible.

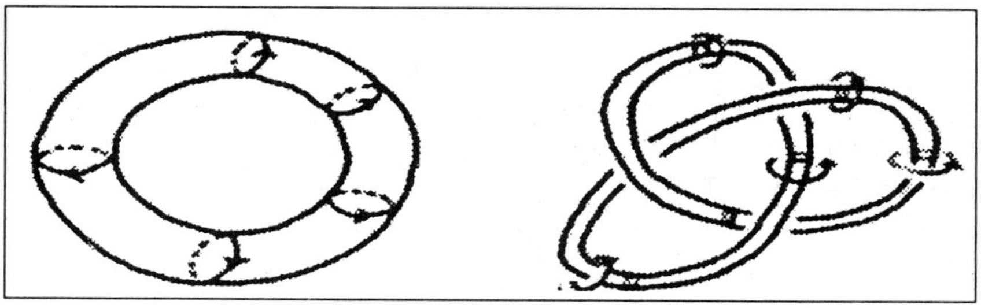

Figure 1. Fluid vortices, also called "water twists."

After reading Helmholtz's paper, the Edinburgh University physicist[1] Peter Guthrie Tait gave a series of lectures on Helmholtz's paper. To demonstrate Helmholtz's result in his lectures, he used an apparatus of his own design that produced vortex rings of smoke. His presentations illustrated quite vividly and dramatically that:

- The vortex rings behaved as independent solids.
- On collision with one another, the vortex rings rebounded as if they were quivering elastic solids, like rings of rubber.
- The smoke rings exhibited fascinating vibration modes about their circular form.
- On each attempt to cut the smoke rings with a knife, the smoke rings would simply wriggle around the knife. The rings were indivisible!

It just so happened that a professional colleague of Tait's, the physicist Sir William Thomson (also later known as Lord Kelvin) was in the audience of one of Tait's lecture-demonstrations. During Tait's presentation, Thomson was struck by the evident permanence and indivisibility of "water twists," as illustrated by Tait's smoke rings. It was at some time into this lecture that Thomson conceived of and created his atomic vortex theory, i.e., that atoms were nothing more than knotted and linked tubular vortices in the then postulated all pervasive fluid called ether. Starting in 1867, Thomson published a series of papers [56] [57] [58] that explained his theory.

[1]In the nineteenth century, physicists were called natural philosophers.

The atomic vortex theory appears to have been well received by the best minds of the nineteenth century. The physicist James Clerk Maxwell was much impressed with Thomson's atomic theory[3] [33]. Maxwell noted that, unlike the other atomic theories of the time, Thomson's theory was based on only a few assumptions. Because of its axiomatic simplicity, it was more likely to represent the physical world than the other existing theories of the time.

Peter Guthrie Tait became so enamored with the theory that he dedicated many years to constructing tables of knots (i.e. ,Tait's "periodical tables"), thus giving birth to the field of knot theory [53].

Tait is also known to have carried on a dialog with Maxwell. In a letter from Maxwell to Tait [33, page 106], Maxwell wrote,

GLENLAIR
DALBEATTIE,
Nov. 13, 1867.

Dear Tait

If you have any spare copies of your translation of Helmholtz on "Water Twists" I should be obliged if you could send me one.

I set [sic] the Helmholtz dogma to the Senate House in '66, and got it very nearly done by some men, completely as to the calculation, nearly as to the interpretation.

Thomson has set himself to spin the chains of destiny out of a fluid plenum as M. Scott set an eminent person to spin ropes from the sea sand, and I saw you had put your calculus in it too. May you both prosper and disentangle your formulae in proportion as you entangle your worbles. But I fear the simplest indivisible whirl is either two embracing worbles or a worble embracing itself.

For a simple closed worble may be easily split and the parts separated

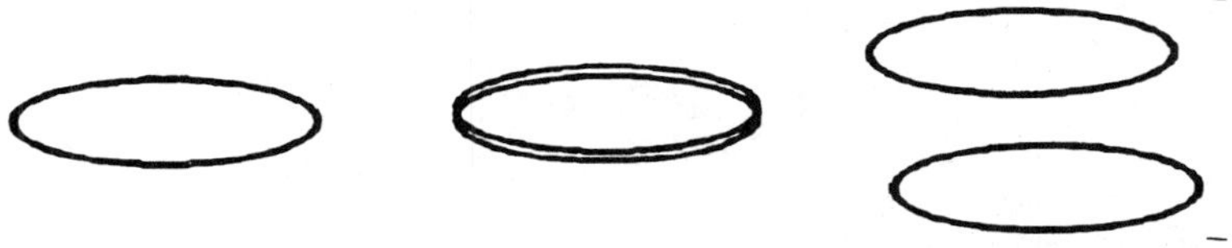

but two embracing worbles preserve each others solidarity thus

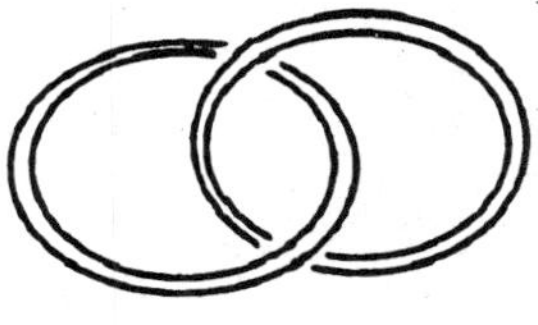

though each may split into many, every one of the one set must embrace every one of the other. So does a knotted one.

yours truly
J. CLERK MAXWELL

Although scientific evidence has since shown conclusively that physical atoms are by no means vortices in the sense of Thomson, Thomson's theory has fragmented and relatively recently reemerged in many much more sophisticated forms in both classical and non classical physics.

In his 1991 Josiah Willard Gibbs Lecture, entitled "The Mysteries of Space," Sir Michael F. Atiyah pointed out four inherent properties of vortex atoms that explain the longevity of vortex theory [3] [4]. These four inherent characteristics are:

- **Stability:** Vortex atoms are stable, as are physical atoms.
- **Variety:** There is a great variety of knots, as there is a great variety of physical atoms.
- **Spectrum:** Vortex atoms have energy states and vibration modes, as do physical atoms.
- **Transmutation:** Knotted vortex atoms change their knot type if their energy is increased beyond a certain threshold, as do physical atoms change their atomic structure.

The recent resurgence of knot theory in physics can be thought of as simply an expression of a recurring major theme in physics, namely:

$$Physics = Geometry$$

Certainly the placement problem is a central part of geometry. A simplistic statement of the placement problem is given below:

The Placement Problem. When are two placements of the same space X in a space Y equivalent?

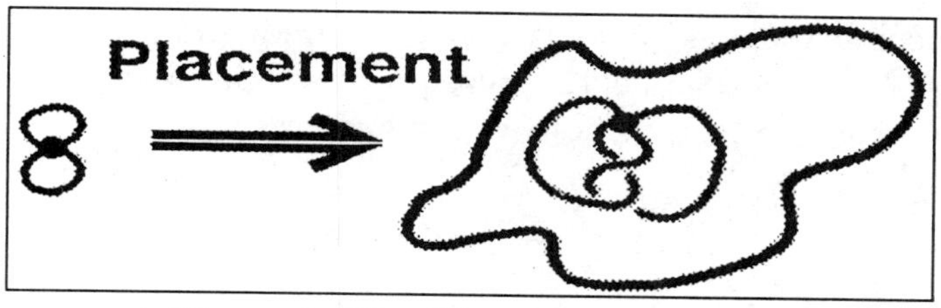

Figure 2. The Placement Problem

2. **Part 1:** KNOTTED MAGNETIC VORTICES IN THE TWENTIETH CENTURY

Our story resumes in the twentieth century with the works of Moffatt, Ricca, Berger, Arnold, Freedman, He, and others [38] [39] [40] [41] [42] [14] [49] [7] [8] [1] [15] [16] [61].

3. PRELIMINARIES: MAGNETIC FIELDS IN PERFECTLY CONDUCTING INCOMPRESSIBLE FLUIDS

Consider an incompressible constant density fluid in $\mathbb{R}^3$ with velocity field $\mathbf{v}(\mathbf{x}, t)$. If the fluid is contained within a closed surface, assume that the velocity field on that surface is tangent to the surface. Because of incompressibility, the velocity field is solenoidal, i.e.,

$$\nabla \cdot \mathbf{v} = 0$$

Let

$$\mathbf{x} \overset{g_t}{\longmapsto} \mathbf{X}(\mathbf{x}, t),$$

denote the fluid flow, where $\mathbf{X}(\mathbf{x}, t)$ denotes the position at time t of a fluid particle that started at position $\mathbf{x}$ at time $t = 0$. Thus, $\mathbf{X}(\mathbf{x}, 0) = \mathbf{x}$. Because of incompressibility, g_t is a parameterized family of volume preserving diffeomorphisms, i.e., $g_t \epsilon SDiff$. Thus, if $\mathcal{R}$ is a region bounded by a smooth surface $\mathcal{S} = \partial\mathcal{R}$, then, as the surface $g_t\mathcal{S}$ (and hence the region $g_t\mathcal{R}$) moves with the flow, the volume

$$\iiint_{g_t \partial \mathcal{R}} dvol$$

remains constant with time. The region $g_t\mathcal{R}$ may of course continuously change its shape. But its volume always remains the same.

Let $\mathbf{B}$ be a magnetic field[2] that is present in the fluid. Since $\mathbf{B}$ is solenoidal, i.e., $\nabla \cdot \mathbf{B} = 0$, there exists a vector potential $\mathbf{A}$ (uniquely determined up to the gradient of an arbitrary scalar function) such that

$$\mathbf{B} = \nabla \times \mathbf{A}$$

[2]With few exceptions, everything said within this paper for a magnetic field $\mathbf{B}$ is equally valid for an arbitrary solenoidal vector field.

Next, assume that the fluid is perfectly conducting. This assumption implies that the magnetic field is frozen in the fluid, and that its behavior is determined by the **frozen field equation**,

$$\frac{\partial \mathbf{B}}{\partial t} = \nabla \times (\mathbf{v} \times \mathbf{B})$$

By 'frozen,' we mean that the magnetic lines of force move with the fluid. Thus, if $\mathcal{S}$ is a surface bounded by a closed curve $\mathcal{C} = \partial\mathcal{S}$, then the magnetic flux Φ crossing the moving surface $g_t\mathcal{S}$, i.e.,

$$\Phi\left(g_t\mathcal{S}\right) = \iint_{g_t\mathcal{S}} \mathbf{B} \cdot d\mathbf{a} = \oint_{g_t\mathcal{S}} A \cdot d\mathbf{x}$$

remains constant with time.

A surface $\mathcal{S}$ is said to be a **magnetic surface** if the magnetic field $\mathbf{B}$ on $\mathcal{S}$ is tangent to $\mathcal{S}$. Since $\mathbf{B}$ is frozen in the fluid, it follows that, if $\mathcal{S}$ is a magnetic surface at time $t = 0$, then the moving surface $g_t\mathcal{S}$ (i.e., the surface moving with the fluid) remains a magnetic surface for all time $t \geq 0$. Moreover, if $\mathcal{S}$ is also a closed surface, then as the parameterized family of magnetic surfaces $g_t\mathcal{S}$ dynamically changes its shape with time t, incompressibility implies that the volume $g_t\mathcal{R}$ which $g_t\mathcal{S}$ encloses

$$\iiint_{g_t\mathcal{R}} d\,vol$$

remains constant. Of course, the magnetic flux leaving a closed magnetic surface is zero.

4. KNOTTED AND LINKED TUBES OF MAGNETIC FLUX

The following definition by the author captures much of the intent and content of the many constructions and examples found within Moffatt's and Berger's papers [39] [40] [41] [42] [7].

Let $\mathbb{T}$ denote the standard solid torus in $\mathbb{R}^3$ given by

$$\left(\quad (2 + \epsilon\cos\theta)\cos\varphi, \quad (2 + \epsilon\cos\theta)\sin\varphi, \quad \epsilon\sin\theta) \quad \right)$$

where $0 \leq \theta, \varphi < 2\pi$ and $0 \leq \epsilon < 1$. For relatively prime integers p and q, let $\mathcal{F}_{p,q}$ denote the foliation of $\mathbb{T}$ by the curves $\gamma_{\epsilon,\theta}$ (where $0 \leq \epsilon \leq 1$ and $0 \leq \theta < 2\pi$) given by

$$\gamma_{\epsilon,\theta}(s) = (\ (2 + \epsilon\cos(\theta + qs))\cos(ps), (2 + \epsilon\cos(\theta + qs))sin(ps), \epsilon\sin(\theta + qs))\ )$$

where $0 \leq s < 2\pi$.

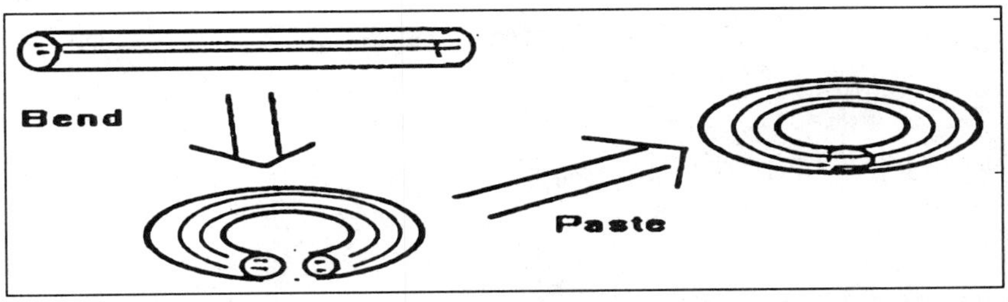

Figure 3. Construction of the foliation $\mathcal{F}_{1,0}$ of the standard torus $\mathbb{T}$.

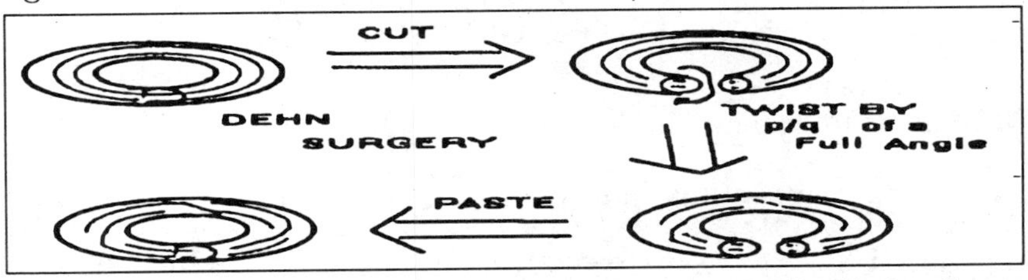

Figure 4. Construction of the foliation $\mathcal{F}_{p,q}$ from $\mathcal{F}_{1,0}$ by Dehn Surgery.

Definition 4.1. [3]*A **magnetic tubular link** (or more briefly, a **magnetic link**) is a smooth immersion into $\mathbb{R}^3$ of finitely many disjoint standard solid tori $\coprod_{i=1}^{n} \mathbb{T}_i$*

$$L : \coprod_{i=1}^{n} \mathbb{T}_i \to \mathbb{R}^3$$

and a smooth magnetic field $\mathbf{B}$ on $\mathbb{R}^3$ such that

(1) *L is an imbedding when restricted to the interior of $\coprod_{i=1}^{n} \mathbb{T}_i$*
(2) *The bounding surface of $\bigcup_i L(\mathbb{T}_i)$, i.e., $\bigcup_i L(\partial \mathbb{T}_i)$ is a magnetic surface*
(3) *For each component $L\mathbb{T}_i$, there exist relatively prime non-zero integers p_i and q_i such that L maps the foliation $\mathcal{F}_{p_i,q_i}$ of $\mathbb{T}_i$ onto the integral curves of $\mathbf{B}$ in $L\mathbb{T}_i$.*

Remark 4.1. *In the above definition, we have allowed L, when restricted to the boundary, to be an immersion to deal with a phenomenon that occurs when the limit is taken under magnetic relaxation of a magnetic link.*

Remark 4.2. *Thus, for every fixed i and j, the linking number between an arbitrary field line in $L\mathbb{T}_i$ and an arbitrary field line in $L\mathbb{T}_j$ is the same regardless of which integral curves are chosen from $L\mathbb{T}_i$ and $L\mathbb{T}_j$ respectively. This is true even when $i = j$.*

[3]This definition encompasses the examples of Moffatt and Berger constructed by Dehn surgery with a twist $2\pi h$ where h is a rational number. To deal with twists for which h is irrational, one should also allow foliations $\mathcal{F}_{p,q}$ where p and q are non-zero real numbers whose ratio p/q is irrational. In this case the curves $\gamma_{\epsilon,\theta}(s)$, $-\infty<s<\infty$, are no longer closed. Arnold's asymptotic linking number [1] would then be required in what is to follow. This is considerably more complicated.

It follows that a magnetic link $\bigcup_i L\mathbb{T}_i$ remains a magnetic link under the action of the fluid flow. I.e., $\bigcup_i g_t L\mathbb{T}_i$ is a magnetic link for $t \geq 0$.

Keeping in mind that the magnetic field $\mathbf{B}$ is frozen in the fluid, our next objective is to find and study those properties of magnetic links that are invariant under the action of fluid flow. One obvious invariant is the **volume** $\mathcal{V}_i$ of each flux tube $g_t L\mathbb{T}_i$, i.e.,

$$\mathcal{V}_i = Vol(L\mathbb{T}_i) = Vol(g_t L\mathbb{T}_i) = \iiint_{g_t L\mathbb{T}_i} dvol$$

which remains unchanged because of incompressibility.

Another invariant of fluid flow is defined as follows:

Definition 4.2. *Let L be a magnetic link. For each solid torus $\mathbb{T}_i$, choose a meridional disk $\mathcal{D}_i$. The* ***magnetic flux*** *$\Phi_i = \Phi(L\mathbb{T}_i)$ in the i-th component is the surface integral defined as*

$$\Phi_i = \Phi(L\mathbb{T}_i) = \iint_{L\mathcal{D}_i} \mathbf{B} \cdot \mathbf{U}\, d\,area$$

where $\mathbf{U}$ denotes the normal to the surface $L\mathcal{D}_i$ pointing in the positive direction induced by the $\mathbf{B}$-field.

It can be shown that Φ_i is independent of the chosen meridional disk. It also can be shown that each Φ_i is a fluid flow invariant, i.e.,

$$\Phi_i(g_t L\mathbb{T}_i) = \iint_{g_t L\mathcal{D}_i} \mathbf{B} \cdot \mathbf{U}\, d\,area$$

is independent of t.

One more fluid flow invariant that will play a central role in the energy minimization of magnetic links is given by the following definition.

Definition 4.3. *The* ***helicity***[4] [5] *of a magnetic link L is defined as:*

$$\mathcal{H}(L) = \iiint_{\bigcup_i L\mathbb{T}_i} \mathbf{A} \cdot \mathbf{B}\, d\,vol$$

---

[4]The term *helicity* was first introduced in a fluid context by Moffatt in [39]. The term *helicity* was previously used in particle physics for the scalar product of the momentum and spin of a particle.

[5]Please note that the *helicity* $\mathcal{H}(L)$ is the same as the *Chern-Simon action.* For $\mathcal{H}(L)$ can be expressed as

$$\mathcal{H}(L) = \int \mathbf{A} \wedge d\mathbf{A} = \int Trace\left(\mathbf{A} \wedge d\mathbf{A} + \frac{2}{3}\mathbf{A} \wedge \mathbf{A} \wedge \mathbf{A}\right)$$

where $\mathbf{A}$ now denotes the magnetic vector potential as a 1-form. For details, see [5].

It can be shown that $\mathcal{H}(L)$ is gauge invariant, and hence well defined.

Theorem 4.1. [41][1] [6] *The helicity is invariant under fluid flow, i.e.,*

$$\frac{d}{dt}\mathcal{H}(g_t L) = 0$$

The following theorem is believed by the author to summarize the many results found within the works of Moffatt, Ricca, and Berger relating the helicity of magnetic links to linking and to magnetic flux [38] [39] [40] [41] [42] [7] [49].

Theorem 4.2. *Let L be a magnetic link. Then*

$$\mathcal{H}(L) = \sum_{1\leq i\leq n} \Phi_i^2 SL_{\mathcal{F}_i} + 2 \sum_{1\leq i<j\leq n} \Phi_i \Phi_j LK_{ij}$$

where $SL_{\mathcal{F}_i}$ denotes the ***self-linking number***[7] *of the axis curve of the tube $L\mathbb{T}_i$ with respect to the framing $\mathcal{F}_i$ induced by the integral curves of the magnetic field $\mathbf{B}$ within $L\mathbb{T}_i$, and LK_{ij} denotes the* ***linking number*** *between any integral curve of the magnetic field $\mathbf{B}$ in $L\mathbb{T}_i$ with any integral curve of the magnetic field $\mathbf{B}$ in $L\mathbb{T}_j$.*

Remark 4.3. *Please note that $SL_{\mathcal{F}_i}$ is the same as the linking number between any two integral curves of the magnetic field $\mathbf{B}$ within the tube $L\mathbb{T}_i$.*

Thus, as noted by Moffatt [38] [39] [40] [41], Berger [7], Arnold [1], Freedman [15] [16], and others [49], the helicity does reflect the topology and the geometry of the magnetic lines of force within a magnetic link.

If for example L has only one component, i.e., L is a magnetic knot, then

$$\mathcal{H}(L) = \Phi^2 SL_{\mathcal{F}}(C)$$

where $SL_{\mathcal{F}}(C)$ is the self-linking number of the axis curve C of the knotted tube with respect to the framing $\mathcal{F}$ induced by the integral curves of the magnetic field $\mathbf{B}$ within the magnetic knot.[8] If for example the tube is knotted in the form of a trefoil and if the magnetic lines of force appear to be parallel to the axis curve when the trefoil is placed on a plane flat surface as shown Fig. 5,

[6] Arnold defines the helicity in a more abstract setting and shows that it is invariant under the group $SDiff$ of volume preserving diffeomorphisms.

[7] For a definition of self-linking number, see for example [22]

[8] This is the same as the linking number between any two distinct magnetic lines of force within the magnetic tube.

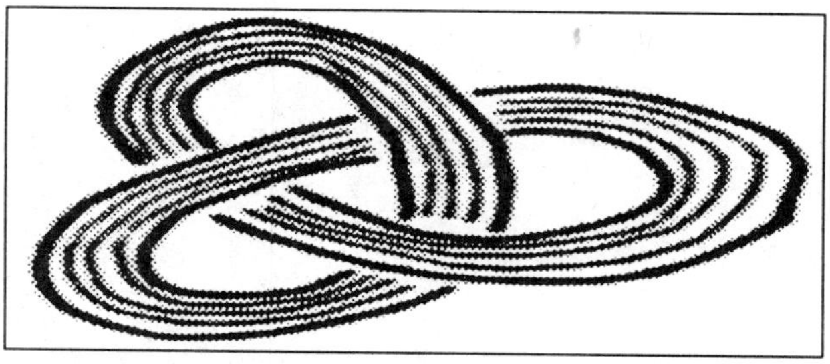

Figure 5. A magnetic trefoil with helicity $\mathcal{H} = \pm 3\Phi^2$.

then $SL = \pm 3$ and

$$\mathcal{H} = \pm 3\Phi^2.$$

On the other hand, if for example the magnetic lines of force induce the trivial framing in each component, then

$$\mathcal{H}(L) = 2 \sum_{1 \le i < j \le n} \Phi_i \Phi_j LK_{ij}$$

Thus, if L is a magnetic two component Hopf link with no twisting of the integral curves of the magnetic field within the components of L as shown in Fig. 6,

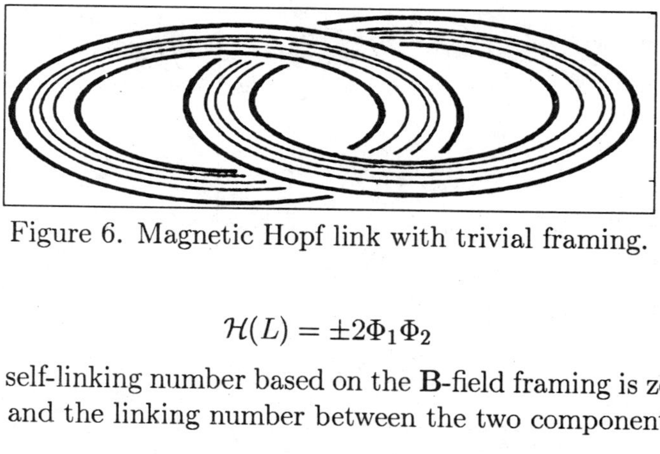

Figure 6. Magnetic Hopf link with trivial framing.

then

$$\mathcal{H}(L) = \pm 2\Phi_1 \Phi_2$$

because the self-linking number based on the **B**-field framing is zero for each component, and the linking number between the two components is ± 1.

Finally, we close this section with the definition of the energy of a magnetic link.

Definition 4.4. *The* ***(magnetic) energy*** $\mathcal{E}_M(L)$ *of a magnetic link* L *is defined by the classical formula*

$$\mathcal{E}_M(L) = \frac{1}{8\pi} \iiint_{\bigcup_i L\mathbb{T}_i} |\mathbf{B}|^2 \, dvol \qquad (\textit{Gaussian units}).$$

Although the energy $\mathcal{E}_M$ is not flow invariant, it will play a central role when magnetic relaxation is discussed in the next section.

5. MAGNETIC RELAXATION AND MINIMUM ENERGY MAGNETIC LINKS

We now turn to the task of studying magnetic relaxation and minimum energy magnetic links.

Consider a magnetic link L in a perfectly conducting, incompressible, viscous fluid. As a result of dissipative frictional fluid forces, the magnetic energy $\mathcal{E}_M(g_tL)$ of g_tL will decrease with time t. In losing energy, the magnetic lines of force will contract. On the other hand, since this is a volume preserving process, the cross sections of the flux tubes of g_tL will at the same time expand. These changes occur while the flux and helicity of g_tL will remain the same.

This process can not continue indefinitely. For eventually the magnetic flux tubes of g_tL must make contact with each other, as indicated in Fig. 7.

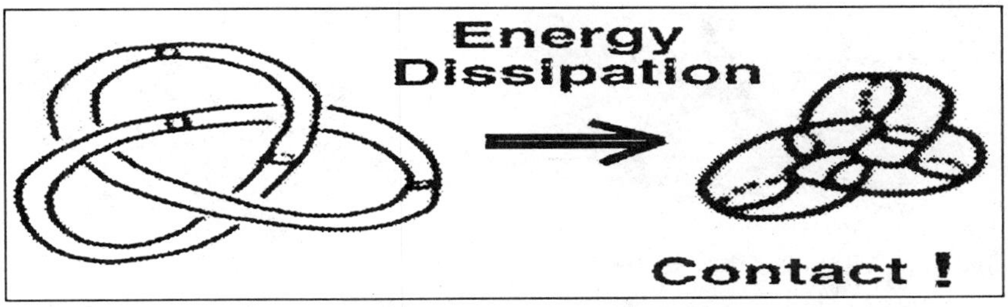

Figure 7. Magnetic relaxation of a magnetic trefoil.

In other words, the topology of the magnetic link g_tL, as expressed in knotting and linking, creates a barrier to the full dissipation of the magnetic link's energy. I.e., $\mathcal{E}_M(g_tL)$ has a positive lower bound that results from the topology of g_tL. Thus, the magnetic link will reach a non-trivial stable and invariant energy state, much as Kelvin conjectured his atomic vortices would.

Since the helicity $\mathcal{H}(g_tL)$ is both an invariant of fluid flow and an expression of the magnetic link g_tL's topology, the following theorem is one mathematical expression of this topological bound.

Theorem 5.1. [40][41] *Let L be a magnetic link. Then*

$$\mathcal{E}_M(L) \geq q_0 \left|\mathcal{H}(L)\right|$$

where q_0 is a non-zero constant that is independent of the magnetic link.

In [15] and [16], Freedman and He obtain more subtle and tighter topological bounds on the minimum energy of magnetic links. For example, for a magnetic knot K they prove that

$$\mathcal{E}_M(K) \geq \frac{1}{4\pi^{\frac{5}{4}}} \frac{\Phi(K)^{\frac{3}{2}} ac(K)^{\frac{3}{4}}}{\mathcal{V}(K)^{\frac{1}{3}}} \geq \frac{1}{4\pi^{\frac{5}{4}}} \frac{\Phi(K)^{\frac{3}{2}} (2g(K)-1))}{\mathcal{V}(K)^{\frac{1}{3}}} \quad (Gaussian\ units)$$

where $\mathcal{V}(K)$ denotes the volume of the magnetic knot K, where $\Phi(K)$ denotes the flux in K, where $ac(K)$ is the asymptotic crossing number, and where $g(K)$ is the genus of the knot K. Freedman and He conjecture that

$$ac(K) = c(K)$$

where $c(K)$ is the crossing number, i.e., the minimum number of crossings among all plane diagrams representing the knot K.

In [40] and [41], Moffatt suggests that the minimum energy spectrum of a magnetic knot can be used to construct new knot invariants.

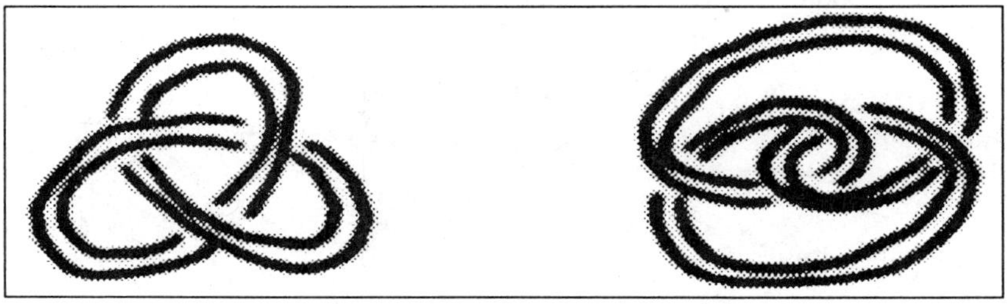

Figure 8. Two different reprsentatives of the trefoil knot type that could possibly magnetically relax to different minimal energy positions. Example taken from [40].

6. **Part 2:** KNOTTED ELECTROSTATIC VORTICES IN THE TWENTIETH CENTURY

A seemingly independent part of our tale of two centuries also resumes in the twentieth century with the works of Freedman, He, Birman, Lomonaco, Kusner, and others [18] [20] [9] [35] [12] [50] [51] [32] [45].

7. PRELIMINARIES: KNOTTED LOOPS AND TUBES OF ELECTRICAL CHARGE AND THE HONEY JAR PROBLEM

Consider a knotted loop α of string or wire that is perfectly flexible but non-extensible, hence of fixed length. Assume that α carries a fixed electrical charge Q. And let α be placed in a non-conducting, non-dielectric viscous fluid which we shall call "honey."

The Honey Jar Problem (HJP)[9]: *Find the minimal energy position or positions eventually assumed by the knotted loop α as a result of the repulsive Coulomb forces in combination with the dissipative frictional fluid forces.*

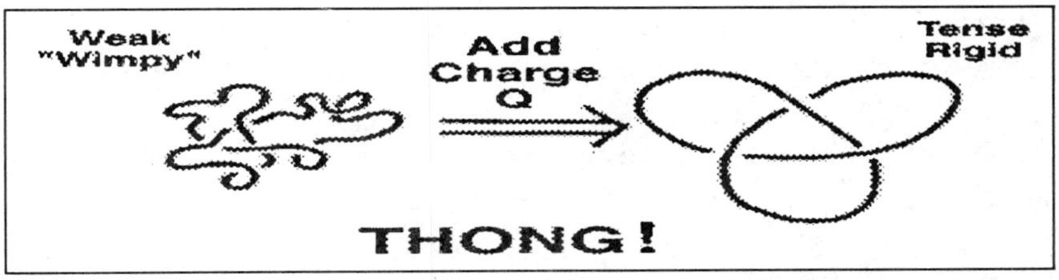

Figure 9. A possible example of an inextensible, flexible knot assuming a minimal energy position when a fixed charge Q is added.

There are actually at least two honey jar problems, i.e.,

- The **conducting honey jar problem** where the charges are allowed to move freely on α, i.e., α is assumed to be a perfect conductor. In this case, we refer to α as a **wire**.
- The **non-conducting honey jar problem** where the charges on α are immobile, i.e., α is assumed to be a perfect non-conductor. In this case, we refer to α as a **string**.

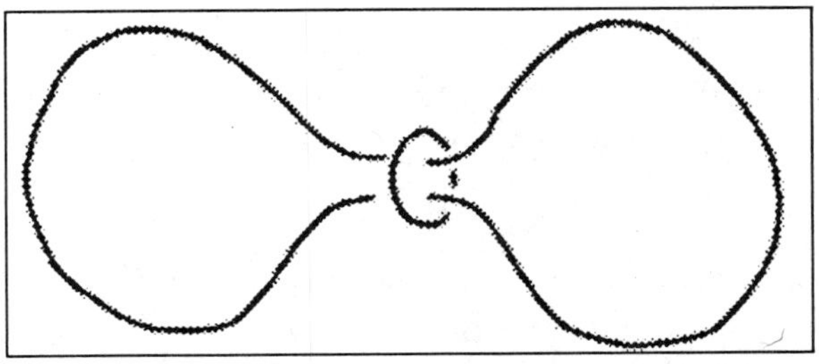

Figure 10. A possible example of a minimal energy electrostatic link.

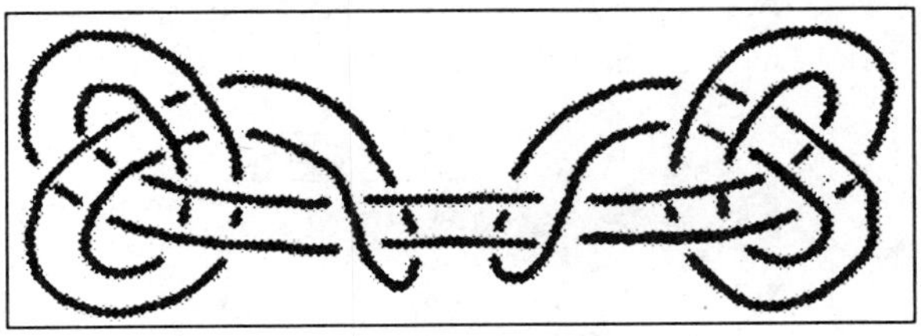

Figure 11. A possible example of a minimal energy electrostatic knot. Example taken from [18].

[9]The phrase 'honey jar' is taken from [20].

In [9] [35], Birman and Lomonaco discussed many of the aspects and difficulties that arise in this problem. In [9], it is pointed out that the loop α can be viewed in at least three ways, thus giving rise to three different honey jar problems for each of the above two cases, namely,

(1) The **HJP for curves** where the loop α is a knotted simple closed space curve in $\mathbb{R}^3$ carrying a charge given by a linear charge density function λ,
(2) The **HJP for hollow tubes** where the loop α is a knotted hollow tube (of constant cross-sectional radius ϵ) in $\mathbb{R}^3$ carrying a surface charge given by a surface charge density function σ,
(3) The **HJP for solid tubes** where the loop α is a knotted solid tube (of constant cross-sectional radius ϵ) in $\mathbb{R}^3$ carrying a charge given by a volume charge density function ρ.

In each of these cases the ultimate goal is to understand physical behavior. Success to this end can best be measured in terms of the following yardstick.

The Asymptotic Behavior Objective (ABO) Yardstick. *The objective is to understand minimum energy asymptotic behavior, i.e., the minimum energy behavior of a physical knotted charged string or wire as its cross section approaches zero.*

8. THE HONEY JAR PROBLEM FOR CURVES: MINIMUM ENERGY KNOTTED SPACE CURVES OF ELECTRICAL CHARGE

One of the major stumbling blocks encountered in the HJP for curves is a theorem in potential theory [29] that states that the potential energy of an electrically charged space curve is necessarily infinite. The potential energy of a simple closed curve $\mathbf{x}(s)$ parameterized by arclength s and having a charge density function $\lambda(s)$ is given by the classical formula

$$\mathcal{E} = \frac{1}{2}\oint\oint\frac{\lambda(s)\lambda(s')dsds}{|\mathbf{x}(s) - \mathbf{x}(s')|}$$

which as stated is divergent. As Jackson points out in [27], the above integral diverges because of "self-energy" contributions of the electric field.

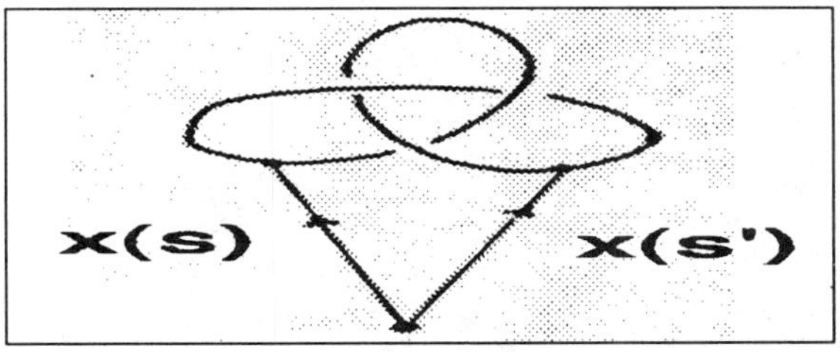

Figure 12. Two position vectors $x(s)$ and $x(s')$ pointing to two distinct points on a trefoil.

This singular behavior (pun intended) calls for great ingenuity for those in pursuit of the minimum energy positions of such curves! A number of renormalization techniques have been employed to navigate around this difficulty.

Fukuhara [21], for example, uses the discrete version of the above integral

$$\frac{1}{2} \sum_{1 \le i < j \le n} \frac{q_i q_j}{|\mathbf{x}_i - \mathbf{x}_j|} \Delta_i s \Delta_j s$$

which is obviously finite. This approach yields physically meaningful results for a discrete set of charges. But are these results meaningful in the limit? I.e., in terms of the ABO yardstick, are these results physically meaningful as each $\Delta_i s$ is approaches zero?

Birman and Lomonaco in [9] and Lomonaco in [35] renormalize by replacing the Coulomb potential of classical electrodynamics $\frac{1}{r}$ with $\frac{1}{r+\epsilon}$, so that the energy $\mathcal{E}_\epsilon$ becomes

$$\mathcal{E}_\epsilon = \frac{1}{2} \oint \oint \frac{\lambda(s)\lambda(s')dsds}{|\mathbf{x}(s) - \mathbf{x}(s')| + \epsilon}$$

It is hoped, but not proven, that a result gained with this definition of energy will give meaningful results about asymptotic behavior, i.e., as ϵ tends to zero.

With this renormalization scheme for the conducting HJP for curves, it can be shown [9] [35] that a minimal ϵ-energy knot with $\mathbf{x}(s)$ C^2 and $\lambda(s)$ C^1 satisfies the following **minimum energy equations**

$$\begin{cases} \nu_\epsilon \frac{d^2\mathbf{x}}{ds^2} & = & -\lambda_\epsilon \mathbf{E}_\epsilon \\ \\ \mu_\epsilon & = & \Phi_\epsilon \end{cases}$$

where the constant ν_ϵ is the **tension** in the curve, where the constant μ_ϵ is the constant **potential** of the conducting curve, where $\mathbf{E}_\epsilon = \mathbf{E}_\epsilon[\mathbf{x}]$ is the ϵ-electric field functional, and where $\Phi_\epsilon = \Phi_\epsilon[\mathbf{x}]$ is the ϵ-potential functional. It is not clear that the solutions $\mathbf{x}_\epsilon$ to the above ϵ-equation approach a limit as ϵ tends to zero. And even if they do, is such a limit physically meaningful?

Although both ν_ϵ and μ_ϵ tend to infinity as ϵ tends to zero, the quantity

$$v_\epsilon + \frac{1}{2}\mu_\epsilon \lambda_\epsilon$$

can be shown to approach a limit. Thus, the limit of the following linear combination of the above ϵ-equations

$$\left(v_\epsilon + \frac{1}{2}\mu_\epsilon\lambda_\epsilon\right)\ddot{\mathbf{x}} = -\lambda_\epsilon \mathbf{E}_\epsilon + \frac{1}{2}\lambda_\epsilon \Phi_\epsilon \ddot{\mathbf{x}}$$

exists.

In terms of the ABO yardstick, are the solutions of the limiting equation physically meaningful?

Freedman, He, and Wang in[18] [20] assume a constant charge density λ of 1, and use a renormalization procedure similar to, but not the same as, one used by O'Hara [45] [46] [47] [48]. Their renormalized energy in [18] [20] for the $\frac{1}{r}$ potential of classical electrodynamics is:

$$\mathcal{E}_1[\mathbf{x}] = \oint\oint \left(\frac{1}{|\mathbf{x}(t) - \mathbf{x}(t')|} - \frac{1}{D(\mathbf{x}(t), \mathbf{x}(t'))} \right) \left| \frac{d\mathbf{x}(t)}{dt} \right| \left| \frac{d\mathbf{x}(t')}{dt'} \right| dt dt'$$

where $D(\mathbf{x}(t), \mathbf{x}(t'))$ denotes the distance between the points $\mathbf{x}(t)$ and $\mathbf{x}(t')$ on the curve $\mathbf{x}(t)$. It is proven that the minimizer of $\mathcal{E}_1$ over the class Γ of all rectifiable, simple closed curves of length 1 is a planar, convex, simple, closed curve.[10] It is also conjectured that any minimizer of Γ is a round circle of circumference 1.

But are the solutions physically meaningful? Does this approach measure up to the ABO yardstick?

In closing this section, it should be noted that there is no infinite energy barrier between knot types for the classical $\frac{1}{r}$-potential energy. I.e., the renormalized $\frac{1}{r}$-potential energy does not blow up as the loop α acquires double points. This is expressed concisely in the equations

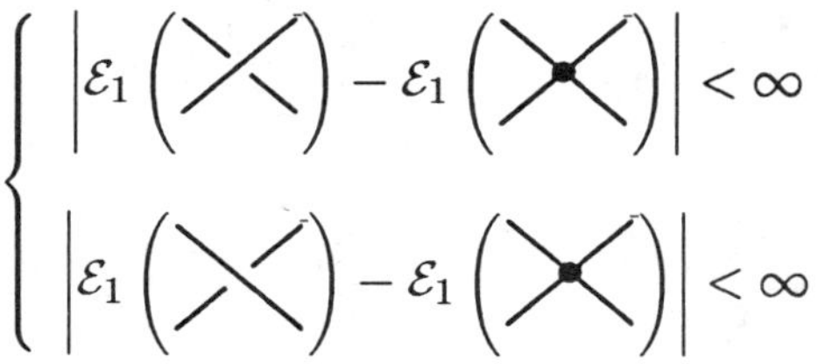

The finiteness of this classical energy barrier fits neatly within Thomson's atomic vortex philosophy. In the last part of the first section of this paper, this is expressed in Atiyah's fourth listed reason for the longevity of Thomson's theory, namely, the *transmutation* characteristic. On the other hand, it does raise a fundamental question that has yet to be resolved. Are there any non-trivial minimal energy knots for the renormalized classical energy? Or in terms of the ABO yardstick, as the cross-section tends to zero, will every charged string or wire try to cross through itself as it seeks a minimal

[10]This does not imply that the minimizer of a representive of any specific knot type is always a planar, convex, simple, closed curve. Please refer to Figures 10 and 11 and to the discussion in the following last three paragraghs of this section.

energy position? Or will it in some cases seek a non-trivial minimal energy position? Figures 10 and 11 indicate possible examples of non-trivial minimal energy knots. Needless to say, it is important to know the answer to this question.

9. A DIGRESSION OUTSIDE THE REALM OF PHYSICS: MÖBIUS "ENERGY" AND OTHER FUNCTIONALS OF KNOTS

We now consider "energies" for knots and links that are based on non-physical potentials.[11] As a result, the ABO yardstick which is the original motivation for the original Honey Jar Problem (HJP) is completely abandoned. (Strictly speaking, this topic does not belong to the purview of this paper because the focus of this manuscript is classical electrodynamics, or at the very least, classical physics.)

The main motivation behind this shift to non-physical "energies" appears to stem mainly from the existence of a finite classical-energy barrier between knot types, as mentioned at the end of the last section. For the purpose of creating an infinite "energy" barrier between knot types, various non-physical "energy" functionals have been devised. For example, an "energy" based on a non-physical $\frac{1}{r^d}$-potential for $d \geq 2$ provides an infinite "energy" barrier between knot types.

Once a non-physical "energy" is chosen, the research objective shifts away from the Honey Jar Problem and from the ABO yardstick objective to the problem of finding the "best" shape, i.e., a shape of minimal "energy", for knots and links. More specifically, the new research objective is to find flows on isotopy classes of knots and links that terminate at a "best" shape for each given knot or link type, and then to use this "best" shape to develop new knot invariants.

One of the most prominent of these "energies" is the "energy" functional based on the non-physical $\frac{1}{r^2}$-potential [21] [45] [46] [47] [48] [32] [43] [12] [18] [20] [34] [13] [50] [51]. Freedman and He's version of this functional, which they call Möbius energy", [18] [20] is defined as

$$\mathcal{E}_2[\mathbf{x}] = \oint\oint \left(\frac{1}{|\mathbf{x}(t) - \mathbf{x}(t')|^2} - \frac{1}{D\left(\mathbf{x}(t), \mathbf{x}(t')\right)^2} \right) \left| \frac{d\mathbf{x}(t)}{dt} \right| \left| \frac{d\mathbf{x}(t')}{dt'} \right| dt dt'$$

where $D\left(\mathbf{x}(t), \mathbf{x}(t')\right)$ denotes the distance between the points $\mathbf{x}(t)$ and $\mathbf{x}(t')$ on the curve $\mathbf{x}(t)$. This functional is independent of the length of the curve $\mathbf{x}(t)$.

[11] Strictly speaking, these "energies" should be more correctly named knot functionals. But we bow begrudgingly to the overwhelming misleading misuse of this term in the mathematical literature.

In [18], Freedman and He prove the surprising result that, $\mathcal{E}_2[\mathbf{x}]$ is invariant under the action of the group $GM(\mathbb{R}^3 \cup \infty)$ of Möbius transformations.[12] More precisely, they prove that for every T in $GM(\mathbb{R}^3 \cup \infty)$

$$\mathcal{E}_2[T\mathbf{x}] = \begin{cases} \mathcal{E}_2[\mathbf{x}] & \text{if } T \text{ takes no point } \mathbf{x} \text{ to } \infty \\ \\ \mathcal{E}_2[\mathbf{x}] - 4 & \text{if } T \text{ takes some point } \mathbf{x} \text{ to } \infty \end{cases}$$

They also show that the topological crossing number $c([\mathbf{x}])$ is bounded by the Möbius "energy," i.e.,

$$2\pi c([\mathbf{x}]) + 4 \leq \mathcal{E}_2[\mathbf{x}]$$

It immediately follows that the number of distinct knot types with representatives below any given Möbius "energy" threshold is bounded. They also prove that a minimal Möbius "energy" representative exists for every irreducible knot type. Moreover, they show that every locally extremal loop for the functional $\mathcal{E}_2$ is $C^{1,1}$.

In [32], Kim and Kusner construct the first explicit example of non-trivially knotted curves which are critical for $\mathcal{E}_2$.

10. THE HONEY JAR PROBLEMS FOR HOLLOW AND SOLID TUBES: MINIMUM ENERGY KNOTTED TUBES OF ELECTRICAL CHARGE

One of the chief difficulties with the HJP for curves is that infinitely thin charged strings and wires simply do not exist.

The HJP problems for hollow and solid tubes are more in keeping with the Kelvin approach. But most importantly, the classical electrostatic energy of charged hollow and solid knotted and linked tubes is a well defined, finite physical quantity [29]. No renormalization is needed.

The classical electrostatic energy of charged hollow tubular links is given by

$$\mathcal{E}_{Surface} = \frac{1}{2} \iint\limits_{\underset{i}{\cup} L\partial\mathbb{T}_i} \iint\limits_{\underset{i}{\cup} L\partial\mathbb{T}_i} \frac{\sigma\sigma'}{|\mathbf{y} - \mathbf{y}'|} da da'$$

where σ, and σ' denote respectively the surface charge densities at points P and P' on the surface, and where P and P' are given respectively by the position vectors $\mathbf{y}$ and $\mathbf{y}'$. The energy $\mathcal{E}_{Solid}$ for a solid tube is defined in like manner.

In [35], Lomonaco studies both the conductor and non-conductor HJP for hollow and solid tubes. For the conductor HJP for hollow tubes, he proves that a minimal energy electrically charged hollow tubular link of constant

[12] For a definition of $GM(\mathbb{R}^3 \cup \infty)$ see [6].

circular cross section of radius ϵ must satisfy the following **minimum energy equations**

$$\begin{cases} \left(v_\epsilon + \mu_\epsilon \frac{Q}{L}\right) \frac{d^2\mathbf{x}}{ds^2} & = & -\oint \sigma_\epsilon \mathbf{E}_\epsilon \omega d\theta \\ \mu_\epsilon & = & \Phi_\epsilon \end{cases}$$

where $\mathbf{x}(s)$ is the center axis curve of the tube parameterized by arclength s, where the constant ν_ϵ is the **tension** in the center curve $\mathbf{x}$, where the constant μ_ϵ is the **electrical potential** of the conducting surface, where σ_ϵ is the surface charge density, where $\mathbf{E}_\epsilon[\mathbf{x}]$ is the electric field functional on the surface, where $\Phi_\epsilon[\mathbf{x}]$ is the electrical potential functional on the surface, and where $\omega d\theta ds$ is the surface area 2-form.

In [35], results are also obtained for the non-conducting HJP for hollow tubes. Moreover, in [35], results for conducting and non-conducting HJP for solid tubes are given.

These approaches are more likely to measure up to the ABO yardstick. But do they?

11. **Conclusion:** CONCLUSIONS FOR THIS CENTURY AND OPEN QUESTIONS FOR THE NEXT MILLENNIUM

In our tale of two centuries, we have traced the impact of Sir William Thomson's (Lord Kelvin's) work on his atomic vortex theory in the nineteenth century to current research in the twentieth century on magnetic and electrostatic knots and links. Clearly, there are many open questions to be answered, especially in regard to dynamic behavior of knotted loops of electrical charge and current, that most certainly will keep researchers occupied much into the next millennium.

REFERENCES

[1] Arnold, V.I., **The asymptotic Hopf invariant and its applications**, Sel. Math. Sov., Vol. 5, No. 4 (1986), pp327-345. (English translation)

[2] Arnold, V.I., **"Mathematical Methods of Classical Mechanics,"** Springer-Verlag, New York (1978).

[3] Atiyah, Sir Michael F., **"The Mysteries of Space,"** The 1991 Josiah Willard Gibbs Lecture, San Francisco, California, January 1991, AMS Selected Lectures in Mathematics (video tape).

[4] Atiyah, Sir Michael F., **"The Geometry and Physics of Knots,"** Cambridge University Press 1990.

[5] Baez, John, and Javier P. Muniain, **"Gauge Fields, Knots and Gravity,"** World Scientific, River Edge, New Jersey (1994).

[6] Beardon, Alan F., **"The Geometry of Discrete Groups,"** Springer-Verlag, New York (1983).

[7] Berger, Mitchell A., and George B. Field, **The topological properties of magnetic helicity**, J. Fluid Mech. (1984), vol. 147, pp 133-148.

[8] Berger, M.A., **Energy-crossing number relations for braided magnetic fields**, Physical Rev. Letters, Vol. 70, No. 6, 8 February 1993, pp 705 - 708.

[9] Birman Joan and S. J. Lomonaco, Jr., **Electrostatic knots,** talk presented at a special session on Low Dimensional Topology at the joint AMS/CMS meeting held in Vancouver, British Columbia, Canada, June 2, 1983.

[10] Brakke, Kenneth A., **"The Motion of a Surface by its Mean Curvature,"** Princeton University Press, Princeton, New Jersey, 1978.

[11] Brakke, Kenneth A., **Surface evolver manual,** Minnesota Geometry Center, Research Report GCG55, July 1993.

[12] Bryson, Steve, Michael H. Freedman, Zheng-Xu He, and Zhenghan Wang, **Möbius invariance of knot energy,** A.M.S. Bull. Vol. 28, No. 1, January 1993, pp. 99 - 103.

[13] Buck, G. and J. Simon, **Knots as dynamical systems,** Topology and its Applications 51 (1993) 229 -246.

[14] Chui, A.Y.K. and H.K. Moffatt, **Minimum energy magnetic fields with toroidal topology,** in "Topological Aspects of the Dynamics of Fluids and Plasmas," ed. H. K. Moffatt, G.M. Xaslavsky, P. Comte, and M. Tabor, Kluwer Academic Publishers (1992), pp 195-218.

[15] Freedman, Michael H., **A note on topology and magnetic energy in incompressible perfectly conducting fluids,** J. Fluid Mech. (1988), vol. 194, pp 549 - 551.

[16] Freedman, Michael H. and Zheng-Xu He, Divergence free fields: **Energy and asymptotic crossing number,** Ann. Math.,134,(1991), 189-229.

[17] Freedman, Michael H. and Zheng-Xu He, **Links of tori and the energy of incompressible flows,** Topology, vol. 30, No. 2 (1991), pp 283 - 287.

[18] Freedman, Michael H., Zheng-Hu He, and Zhenghan Wang, **Möbius Energy of Knots and Unknots,** Ann. Math.,139 (1994),I-50.

[19] Freedman, M.H. and F. Luo, **Equivariant isotopy of unknots to round circles.** Preprint submitted to Topology and Its Applications.

[20] Freedman, Michael H. and Zheng-Xu He, **Research announcement on the "energy" of knots,** in "Topological Aspects of the Dynamics of Fluids and Plasmas," ed. H.K. Moffatt, G.M. Xaslavsky, P. Comte, and M. Tabor, Kluwer Academic Publishers (1992), pp 219 - 222.

[21] Fukuhara, Shinji, **Energy of a knot,** pp 443-451, in "A Fete of Topology: Papers Dedicated to Itiro Tamura," edited by Matsumoto,Y.T., and S. Morita, Academic Press, New York (1988).

[22] Fuller, F. Brock, **The writhing number of a space curve,** Proc. Nat'l Acad. Sci. USA, Vol. 68, No. 4, 1971, pp 815 - 819.

[23] Gauss, Karl Friedrich, **Gauss 1833 Werke. Königlichen Gesellschaft der Wissenschaften zu Göttingen,** 1877, p 605 -

[24] Goldstein, Herbert, **"Classical Mechanics,"** (second edition) Addison-Wesley, Reading, Massachusetts (1980).

[25] Green, George and John T. Lloyd (editors), **"Kelvin Papers: Index to the manuscript collection of William Thomson, Baron Kelvin in Glasgow University Library,"** Glasgow University Library (1977).

[26] Helmholtz, Herman von, **Über Integrale, der hydrodynamischen Gleichungen, welche den Wirbelbewegungen entsprechen,** Jour. Reine Angewandte Mathematik, 55 (1858), pp 25 - 55.

[27] Jackson, John David, **"Classical Electrodynamics,"** (second edition) John Wiley & Sons, New York (1975).

[28] Kauffman, Louis H., **"Knots and Physics"** (second Edition), World Scientific, New Jersey (1993).

[29] Kellogg, Oliver Dimon, **"Foundations of Potential Theory,"** Dover, New York (1954).

[30] Kelvin, Lord (also known as Sir William Thomson), **"Baltimore Lectures on Molecular Dynamics and the Theory of Light,"** C.J. Clay and Sons, Cambridge University Press Warehouse, Ave Maria Lane, London (1904).

[31] Kelvin, Lord (also known as Sir William Thomson), and Peter Guthrie Tait, **"Treatise on Natural Philosophy,"** Volumes I & 2, Cambridge University Press (1923).

[32] Kim, Denise and Rob Kusner, **Torus knots extremizing the Möbius energy**, Experimental Math, 2, (1993), pp I - 9.

[33] Knott, Cargill Gilston, **"Life and Scientific Work of Peter Guthrie Tait,"** Cambridge University Press (1911).

[34] Kusner, Robert B., and John M. Sullivan, **Möbius energies for knots and links, surfaces and manifolds**, Minnesota Geometry Center, Research Report GCG64, University of Minnesota, Minneapolis, MN.

[35] Lomonaco, S.J., Jr., **Minimum energy knotted tubes of electrical charge**, in preparation.

[36] Maxwell, James Clerk, **On physical lines of force**, Parts I-11, March, April, May 1861, Parts III-IV, Jan., Feb., 1862.

[37] Moffatt, H.K., **"Magnetic Field Generation in Electrically Conducting Fluids,"** Cambridge University Press, New York, New York (1978).

[38] Moffatt, H.K., **The degree of knottedness of tangled vortex lines**, J. Fluid Mech., vol. 35, part 1 (1969) pp. 117 -129.

[39] Moffatt, H.K., **Magnetostatic equilibria and analogous Euler flows of arbitrarily complex topology. Part 1. Fundamentals**, J. Fluid Mech. (1985), vol. 159, pp. 359 -378.

[40] Moffatt, H.K., **The energy spectrum of knots**, Nature, vol. 347, September 1990, pp 367 -369.

[41] Moffatt, H.K., **Relaxation under topological constraints**, in "Topological Aspects of the Dynamics of Fluids and Plasmas," ed. H.K. Moffatt, G.M. Xaslavsky, P. Comte, and M. Tabor, Kluwer Academic Publishers (1992), pp 3-28.

[42] Moffatt, H.K. and Renzlo L. Ricca, **Helicity and the Calugareanu invariant**, Proc. R. Soc. London A (1992) 439, pp 411-429.

[43] Nakauchi, Nobumitsu, **A remark on O'Hara's energy of knots**, Proc. AMS, vol. 118, No. 1, May 1993, pp 293 - 296.

[44] Niven, W.D. (editor), **"The Scientific Papers of James Clerk Maxwell,"** Dover, New York, 1966.

[45] O'Hara, Jun, **Energy of a knot**, Topology, vol. 30, No.2, pp241 -247, 1991.

[46] O'Hara, Jun, **Energy functional of knots**, "Topology Hawaii," K.H. Dovernan ed., World Scientific.

[47] O'Hara, Jun, **Energy functionals of knots**, Topology and its Applications.Ohara4

[48] O'Hara, Jun, **Energy functionals of knots II**, preprint

[49] Ricca, R. L., and H.K. Moffatt, **The helicity of a knotted vortex filament**, in "Topological Aspects of the Dynamics of Fluids and Plasmas," ed. H.K. Moffatt, G.M. Xaslavsky, P. Comte, and M. Tabor, Kluwer Academic Publishers (1992), pp 225 -236.

[50] Simon, J., **Energy functions of knots**, preprint.

[51] Simon, Jonathan K., **Energy functions of polygonal knots**, J. Knot Theory and Its Ramifications, vol. 3, no. 3 (1994) pp 299-320.

[52] Stewart, Ian, **Finding the energy to solve a knotty problem**, Science, 6 March, 1993, p. 18.

[53] Tait, P.G., **"On Knots 1, 11, III,"** Scientific Papers, Vol. I, Cambridge University Press, London (1898), pp273-347.

[54] Tait, P.G., **Scientific Papers**, Vol. 11, Cambridge University Press (1900).

[55] Thomson, J.J., **"Notes on Recent Researches in Electricity and Magnetism,"** Clarendon Press, Oxford (1893).

[56] Thomson, W.T. (Lord Kelvin), **On vortex atoms**, Philosophical Magazine, 34 July 1867, p15-24. Mathematical and Physical Papers, Vol.4, Cambridge (1910).

[57] Thomson, Sir William (Lord Kelvin), **On vortex motion**, Trans. Royal Soc. Edinburgh, 25 (1868/69), pp 217-260.

[58] Thomson, Sir William (Lord Kelvin), **Vortex statics**, Proceedings of the Royal Society of Edinburgh, Vol. IX , No. 94 (1875), pp 59 - 78.

[59] Thomson, Sir William (Lord Kelvin), **"Reprints of Papers on Electrostatics and Magnetism,"** (second edition) Macmillan & Co., London (1884).

[60] Wilson, David B. (editor), **"Catalogue of the Manuscript Collections of Sir George Stokes and Sir William Thomson, Baron Kelvin of Largs in the Cambridge University Library,"** Cambridge University Library (1976).

[61] Woltjer, L., **A theorem on force-free magnetic fields**, Proc. Nat'l Acad. Sci. U.S.A., **44** (1958), pp 489 - 491.

The author writing

The author lecturing

University of Maryland Baltimore County, Baltimore, Maryland 21228
E-mail address: `Lomonaco@UMBC.Edu`

Proceedings of Symposia in Applied Mathematics
Volume **51**, 1996

SPIN NETWORKS IN NONPERTURBATIVE QUANTUM GRAVITY

JOHN C. BAEZ

ABSTRACT. A spin network is a generalization of a knot or link: a graph embedded in space, with edges labelled by representations of a Lie group, and vertices labelled by intertwining operators. Such objects play an important role in 3-dimensional topological quantum field theory, functional integration on the space $\mathcal{A}/\mathcal{G}$ of connections modulo gauge transformations, and the loop representation of quantum gravity. Here, after an introduction to the basic ideas of nonperturbative canonical quantum gravity, we review a rigorous approach to functional integration on $\mathcal{A}/\mathcal{G}$ in which $L^2(\mathcal{A}/\mathcal{G})$ is spanned by states labelled by spin networks. Then we explain the 'new variables' for general relativity in 4-dimensional spacetime and describe how canonical quantization of gravity in this formalism leads to interesting applications of these spin network states.

1. INTRODUCTION

Spin networks are a generalization of knots and links, and in what follows we would like to describe some of their recent applications to quantum gravity, and also a little bit of their role in knot theory. Knot theory is a well-established branch of mathematics, full of interesting theorems, and it is easy as a mathematician to acquaint oneself with it by reading any of a number of texts. Quantum gravity, on the other hand, is more difficult for the mathematician, consisting as it does mainly of unsolved and usually imprecisely posed questions. Thus it seems worthwhile to start with a thumbnail sketch of quantum gravity, focusing on an approach known as 'nonperturbative canonical quantization'.

After this heuristic introduction, we give a mathematically rigorous account of spin networks in Section 2. For us, spin networks will be graphs embedded in a manifold S, with edges labelled by representations of a Lie group G and with vertices labelled by intertwining

1991 *Mathematics Subject Classification.* Primary 28C20, 81V17, 83C27, 83C45, 83E05.

operators. Spin networks define gauge-invariant functions on the space $\mathcal{A}$ of connections on any G-bundle over S, and we shall show that these functions span the gauge-invariant subspace $L^2(\mathcal{A}/\mathcal{G})$ of a certain Hilbert space $L^2(\mathcal{A})$. In Sections 3 and 4 we go back to the physics and give a more detailed description of the role spin networks play in quantum gravity.

In the latter part of the twentieth century, much energy has been spent in a struggle to reconcile two brilliant accomplishments of the early part of the century: general relativity and quantum theory. So far there is remarkably little to show for all this work when it comes to verified predictions of experimental results. Indeed, it is quite possible that the new predictions of a theory of quantum gravity can only be tested at extremely small length scales, far below those that can be probed by current experimental techniques. The reason for this is simple dimensional analysis: from Planck's constant $\hbar$, Newton's gravitational constant κ, and the speed of light c one can form a quantity with units of length, the Planck length

$$\ell_P = (\hbar\kappa/c^3)^{1/2},$$

in a unique way (up to dimensionless constant factors). This works out to be about 10^{-35} meters. If we are hoping to get experimental evidence for any theory of quantum gravity in the forseeable future, we have to hope that somehow this simple argument is wrong. There are certainly many phenomena already observed, but so far unexplained, that a theory of quantum gravity might hope to 'retrodict': phenomena from particle physics, phenomena from cosmology, even phenomena such as the 4-dimensionality of spacetime that are taken for granted in all current theories of physics. Physicists have suggested many ideas along these lines, but none command widespread acceptance at this time; they are all somewhere between controversial theories and sheer lunacy.

In short, if there were several competing well-established theories of quantum gravity, the lack of empirical evidence to decide between them could easily be a serious problem. Luckily nature has been kind to us, in that we have been unable to formulate *even one* theory combining general relativity and quantum theory in a manner that is *mathematically consistent, not in obvious contradiction with experiment,* and *elegant.* The importance of mathematical consistency and non-contradiction with experiment should be obvious, but the role of elegance deserves some comment. First, it is always possible to find infinitely many consistent theories that fit a given finite set of experimental results, just as there are infinitely many curves through a finite set of points, and without the freedom to reject most of them as appallingly ugly, we would never get anywhere. Second, there is the fact that taken separately, general relativity and quantum theory are strikingly esthetic as pure mathematics. This could be a mere coincidence, but in the absence of evidence to the contrary it is probably best to take it as a hint, and search for a theory combining the most beautiful aspects of both in an integral manner. Indeed, though the 'unreasonable effectiveness' of elegant mathematics is still a great mystery, it is also an empirical fact, so even as hard-nosed empiricists we should heed it.

Many physicists, particularly in string theory, have argued that it is impossible to reconcile general relativity and quantum theory without taking all the other forces and particles into account. Here however we concentrate on a diametrically opposite approach, which tries to construct a quantum version of the *vacuum* Einstein equations — the theory of gravity with *no* matter around. We shall not attempt to discuss the relative merits of the two approaches, but simply note the curious fact that some of the same mathematics shows up in both, possibly for some deep reasons that would be good to understand [11].

In what follows we need to assume a nodding acquaintance with some differential geometry. Let us take as our spacetime a 4-manifold M diffeomorphic to $\mathbb{R} \times S$, where $\mathbb{R}$ represents time and the 3-manifold S (compact and orientable to simplify the discussion) represents space. The vacuum Einstein equation concerns a Lorentzian metric g on M. Given such a metric there is a unique metric-preserving, torsion-free connection Γ, the Levi-Civita connection. The curvature of this connection is described by the Riemann tensor R, which in local coordinates is written $R^\alpha{}_{\beta\gamma\delta}$. A certain amount of information about the curvature of spacetime is thus contained in the Ricci tensor $R_{\mu\nu} = R^\alpha{}_{\mu\alpha\nu}$, and the vacuum Einstein equation says simply that

$$R_{\mu\nu} = 0.$$

To understand why precisely *these* are the equations describing the curvature of empty spacetime, though, it is necessary to step back and consider the case where there is matter present. In the presence of matter, the density of energy and momentum and their rate of flow in different spatial directions are summarized by a symmetric tensor $T_{\mu\nu}$. The sense in which energy and momentum are conserved in general relativity is quite subtle, but the simplest way of stating this conservation principle is that $T_{\mu\nu}$ is 'divergence-free':

$$\nabla^\mu T_{\mu\nu} = 0,$$

where ∇ is the operator of covariant differentiation corresponding to the Levi-Civita connection. If we wish to say that spacetime is curved by energy and momentum, it is natural to try some equation like

$$C_{\mu\nu} = T_{\mu\nu}$$

where C is built up from the Riemann tensor in some simple way. However, are constrained by the fact that C must be symmetric and divergence-free. The simplest thing to try is some multiple of

$$G_{\mu\nu} = R_{\mu\nu} - \frac{1}{2}R^{\alpha}_{\alpha}g_{\mu\nu},$$

because this is symmetric and, by the Bianchi identity, also divergence-free. In fact, Einstein's equation in the presence of matter says that

$$G_{\mu\nu} = 8\pi\kappa T_{\mu\nu}.$$

But in a vacuum $T_{\mu\nu} = 0$, so we obtain

$$G_{\mu\nu} = 0,$$

which turns out to be equivalent to the vanishing of the Ricci tensor.

To see how the vacuum Einstein equation describes the *dynamics* of the metric, it is useful to split spacetime into space and time. The manifold M is diffeomorphic to $\mathbb{R} \times S$, but not in any canonical way. Nonetheless, let us go ahead and pick a diffeomorphism between them. This lets us transfer the usual coordinate function on $\mathbb{R}$ to a 'time' function x^0 on M, so we can talk about the metric on space at time t, that is, the metric g restricted to the slice

$$S_t = \{x^0 = t\} \subseteq M.$$

Assume we can choose the slices S_t to be spacelike, meaning that the restriction of g to each one is a Riemannian metric. We can only do this if g is 'globally hyperbolic', but physically this is a reasonable condition. Then, near any point $p \in M$, we can find local coordinates x^μ such that x^0 is the above time function, the vector field ∂_0 is normal to the slices, and the vector fields ∂_i corresponding to the remaining 'space' coordinates x^i $(i = 1, 2, 3)$ are tangent to the slices. One does not really need to use coordinates with these special properties, but it simplifies the discussion below.

Since $G_{\mu\nu}$ is a symmetric 4×4 matrix, the vacuum Einstein equation really has 10 components. Splitting spacetime into space and time lets us interpret these components as follows. The equation $G_{00} = 0$ and the 3 equations $G_{0i} = 0$ are *constraints* on the metric on space, g_{ij}, and its first time derivative $\dot{g}_{ij}$. In other words, G_{00} and G_{0i} can be computed on the slice S_t in terms of the 'initial data' g_{ij} and $\dot{g}_{ij}$ on that slice, and for there to be a solution of Einstein's equation having given initial data, the initial data must satisfy the constraints $G_{00} = G_{0i} = 0$. The remaining 6 equations $G_{ij} = 0$ are *evolutionary equations* involving the second time derivative $\ddot{g}_{ij}$. They allow us to compute how the metric evolves in time, given that the constraints hold initially.

In fact, the constraints have a simple meaning, which turns out to be very important. To understand this meaning, however, we need a brief detour into classical mechanics. It is worth recalling the simplest classical mechanics problem of all, the motion of a particle on the line in a potential V. The particle's position is a point q in the space $\mathbb{R}$, and one calls this space of possible positions the 'configuration space'. The motion of the particle is determined by its position q and velocity $\dot{q}$ at time zero, but it is often handier to work not with its velocity but with its momentum $p = m\dot{q}$, where m is its mass. The position and momentum determine a point (q, p) in the 'phase space' $T^*\mathbb{R}$, since the momentum is most naturally regarded as a cotangent vector to the configuration space. This phase space has a closed nondegenerate 2-form, or 'symplectic structure', on it, given by

$$\omega = dp \wedge dq.$$

The energy, or Hamiltonian, of the particle is a function on phase space, given by the sum of kinetic and potential energy:

$$H = \frac{p^2}{2m} + V(q).$$

The Hamiltonian gives rise to a vector field v_H on phase space by

$$\omega(\cdot, v_H) = dH.$$

This vector field generates a one-parameter group of diffeomorphisms of phase space, or 'flow', which describes the time evolution of the particle. For short, one says that the Hamiltonian generates time evolution. Indeed, quite often in classical mechanics the phase space is a symplectic manifold and time evolution is generated by the Hamiltonian in this manner. And quite often the phase space is the cotangent bundle of some configuration space; cotangent bundles are equipped with a canonical symplectic structure.

In general relativity, the quantity analogous to the 'position' is the metric g_{ij} on space, and we shall henceforth write this as q_{ij} to emphasize the analogy. The configuration space of general relativity is thus the space $\mathcal{M}$ of all Riemannian metrics on S. The quantity analogous to the 'velocity' is the time derivative $\dot{q}_{ij}$. Following certain standard recipes, the analog of the momentum works out to be

$$p^{ij} = \frac{1}{2}(\det q_{ij})^{1/2}(\dot{q}^{ij} - \dot{q}^k_k q^{ij}).$$

The phase space of general relativity is the cotangent bundle $T^*\mathcal{M}$, and the pair (q_{ij}, p^{ij}) determines a point in this phase space.

Since one can write the constraints G_{00} and G_{0i} in terms of q_{ij} and p^{ij}, one can think of these constraints as functions on $T^*\mathcal{M}$. From this point of view, they play a dual role [27]. First, as already noted, for the pair (q_{ij}, p^{ij}) to determine a solution of Einstein's equation, it must lie on the subspace of $T^*\mathcal{M}$ where the constraints vanish. Second, and more subtly, the constraints generate physically important flows on phase space.

For example, if we integrate G_{00} over S, we obtain a function on $T^*\mathcal{M}$ which generates time evolution with respect to the time coordinate x^0. For this reason physicists call this constraint the 'Hamiltonian constraint', and use the notation

$$\mathcal{H} = G_{00}$$

in this context. On the other hand, if we take a vector field N on S and integrate $N^i G_{0i}$ over S, we obtain a function on $T^*\mathcal{M}$ generating a flow which is the same as that induced by the one-parameter group of diffeomorphisms of S generated by N. Physicists call this constraint the 'diffeomorphism constraint', and write it as

$$\mathcal{H}_i = G_{0i}.$$

It is important to note, however, that both $\mathcal{H}$ and $\mathcal{H}_i$ correspond to one-parameter groups of diffeomorphisms of spacetime: $\mathcal{H}$ corresponds to diffeomorphisms that 'push S forwards in time', while $\mathcal{H}_i$ corresponds to diffeomorphisms that map S to itself.

Now let us turn the problem of quantizing general relativity, that is, guessing a quantum theory that reduces to Einstein's equation in the limit $\hbar \to 0$. Again, it is good to recall the example of a particle on the line. In the quantum theory of a particle on a line, the particle's state is given by a 'wavefunction', that is, an L^2 function ψ on the configuration space $\mathbb{R}$. If we assume ψ is normalized so that $\|\psi\| = 1$, the probability of finding the particle in any open subset $U \subseteq \mathbb{R}$ is then given by

$$\int_U |\psi(x)|^2 \, dx.$$

Now, classically the Hamiltonian is

$$H(p,q) = \frac{p^2}{2m} + V(q),$$

and to quantize the particle one simply replaces p and q in this formula with certain operators $\hat{p}, \hat{q}$ on $L^2(\mathbb{R})$:

$$(\hat{p}\psi)(x) = \frac{\hbar}{i}\frac{d}{dx}\psi(x), \qquad (\hat{q}\psi)(x) = x\psi(x),$$

obtaining an operator $\hat{H}$, the quantum Hamiltonian:

$$(\hat{H}\psi)(x) = -\frac{\hbar^2}{2m}\frac{d^2}{dx^2}\psi(x) + V(x)\psi(x).$$

The Hamiltonian $\hat{H}$ then describes the evolution of the wavefunction in time via Schrödinger's equation:

$$i\hbar\frac{d}{dt}\psi = \hat{H}\psi.$$

This recipe for replacing p and q by operators $\hat{p}$ and $\hat{q}$ is known as 'canonical quantization'. If we try to copy this recipe in the case of general relativity, we quickly notice some serious problems. In the case of the particle on the line, the 'configuration space' in which q lies is just the real line, and we can define $L^2(\mathbb{R})$ using Lebesgue measure. In the case of general relativity, the configuration space is the space $\mathcal{M}$ of all Riemannian metrics on S. This is an *infinite-dimensional* manifold, and there is no good notion of 'Lebesgue measure' on such spaces, so defining the Hilbert space $L^2(\mathcal{M})$ is not so easy.

Of course, this problem will arise whenever we try to canonically quantize a theory with infinite-dimensional configuration space. Thus it occurs, not only in quantum gravity, but in other quantum field theories. When quantizing linear equations, where the configuration space is an infinite-dimensional vector space $\mathcal{V}$, one can use an infinite-dimensional analog of a Gaussian measure to define $L^2(\mathcal{V})$, and the resulting theories make fine mathematical sense. This strategy is not sufficient by itself to deal with nonlinear equations such as Einstein's equation, however. Indeed, $\mathcal{M}$ is not even a vector space.

Another problem is that unlike the particle on the line, whose position and velocity at a given time are arbitrary, general relativity involves *constraints.* One should take these constraints into account when quantizing the theory, but the correct way to do so is not at all clear! Indeed, much ink has been spilled concerning the quantization of constrained systems, and we cannot go into all the nuances of this issue here. Instead, let us simply give an oversimplified account of Dirac's approach to constraints [24], as applied to gravity by DeWitt [23] in the late 1960s.

Suppose we could somehow get ahold of a Hilbert space $L^2(\mathcal{M})$. The idea is that not all vectors in this so-called 'kinematical' state space represent physical states of quantum gravity, but only those satisfying certain quantum versions of the constraints. To describe these quantized constraints, first we try to define operators $\hat{q}_{ij}$ and $\hat{p}^{ij}$ on $L^2(\mathcal{M})$ by the formulas

$$(\hat{p}^{ij}\psi)(q) = \frac{\hbar}{i}\frac{\partial}{\partial q_{ij}}\psi(q), \qquad (\hat{q}_{ij}\psi)(q) = q_{ij}\psi(q),$$

where $q \in \mathcal{M}$. These formulas are merely heuristic, and part of the problem is making good enough sense of them for the task at hand. For example, in analogy with p^{ij} and q_{ij} in the first place, $\hat{p}^{ij}$ and $\hat{q}_{ij}$ are really something like operator-valued functions on the subset of the spacelike slice contained in our coordinate chart. When one tries to be rigorous one discovers that they are operator-valued distributions. As noted above, one can work out formulas for the constraints $\mathcal{H}$ and $\mathcal{H}_i$ in terms of p^{ij} and q_{ij}. The idea is then to take these formulas, and replace all appearances of p^{ij} and q_{ij} in them by $\hat{p}^{ij}$ and $\hat{q}_{ij}$, obtaining operator-valued distributions $\hat{\mathcal{H}}$ and $\hat{\mathcal{H}}_i$. For a wavefunction $\psi \in L^2(\mathcal{M})$ to represent a physical state of quantum gravity, the following quantum versions of the constraint equations should hold:

$$\hat{\mathcal{H}}\psi = 0, \qquad \hat{\mathcal{H}}_i\psi = 0.$$

There are some deep conceptual problems associated with this approach to quantum gravity. For example, instead of Schrödinger's equation, in which the Hamiltonian describes the time evolution of the wavefunction, all we have is the so-called 'Wheeler-DeWitt equation' $\hat{\mathcal{H}}\psi = 0$. What does this mean? Where has the *dynamics* of quantum gravity gone, if physical states are annihilated by the operator whose classical counterpart generates time evolution? Briefly, the answer appears to be this: the constraint equations say that ψ only describes *diffeomorphism-invariant* information about the world. In other words, we cannot use ψ to answer questions about what is happening at a certain point whose location is specified using a coordinate system, such as 'what is the metric at the point $x^\mu = a^\mu$?'. Unfortunately, we are not used to doing physics without asking such questions! This is known as the 'problem of time' in quantum gravity [33].

Just as bad as the conceptual problems are the sheer technical problems involved in making mathematical sense out of DeWitt's strategy for quantizing gravity. In fact, nobody has yet succeeded. It is worth noting that there are other constrained nonlinear equations with infinite-dimensional configuration spaces where many of the same problems show up: for example, the Yang-Mills equation, various versions of which appear to describe all the forces *except* gravity. Quantizing these in a rigorous way is *also* extremely difficult, but physicists have had much more success with them at the practical level. Let us briefly summarize the difference between the two cases.

Physicists often quantize nonlinear field theories by treating the nonlinearities as 'small perturbations' of some linear equation. There are also perturbative methods for dealing with constraints. These methods can be problematic, mathematically speaking, but unless afflicted by an unusual desire for rigor, physicists are often happy to do perturbation theory using formal power series in some coupling constant that measures the nonlinearity. Due to the nonrigorous nature of these computations, the coefficients of these power series usually come out to be ill-defined unless one performs some clever maneuvers known as renormalization. One says the theory is renormalizable if these maneuvers work, but renormalizability by itself does not mean that the power series are convergent, or even asymptotic. Thus, strictly speaking, renormalizability is not a sufficient condition for a nonlinear quantum field theory to make rigorous sense. Nor, on the other hand, is renormalizability a *necessary* condition to be able to quantize a nonlinear wave equation, for there are also other 'nonperturbative' approaches. In practice, however, particle physicists often restrict themselves to renormalizable theories, make physical predictions using the first few terms in the power series, and compare these predictions with experiment to see if the theory is on the right track.

For the Yang-Mills equation, and indeed for the whole 'Standard Model' of particles and forces other than gravity, this strategy has been quite successful. But the strategy fails miserably with quantum gravity, for the theory resists all attempts at renormalization. As it turns out, this is closely related to the fact that the constraints are not polynomials in the basic 'position' and 'momentum' variables, q_{ij} and p^{ij}, and their derivatives. Instead, they contain nasty factors of $(\det q_{ij})^{-1/2}$, essentially because the formula for p^{ij} contains a factor of $(\det q_{ij})^{1/2}$. Standard methods for replacing q_{ij} with the operator $\hat{q}_{ij}$ lead to all sorts of problems when applied to such non-polynomial expressions. (Similar problems arise in the path-integral approach which is more commonly used in renormalization.) To overcome the nonrenormalizability of quantum gravity, particle physicists constructed ever more complicated models containing gravity and other forces, in order to cancel out unruly infinities. This led to the study of supergravity and eventually superstring theory, which attempts to model *all* known forces and particles, and unfortunately many unknown ones as well.

Meanwhile, many general relativists were suspicious of the whole idea of perturbatively quantizing gravity. This amounts to treating all metrics as small perturbations of a fixed 'background metric', usually taken to be the flat Minkowski metric on $\mathbb{R}^4$. Dimensional analysis, however, suggests that one should expect more and more extreme quantum fluctuations of the metric at smaller and smaller length scales, becoming very significant at about the Planck length. So perhaps one should really adopt some *nonperturbative* approach to canonical quantum gravity.

Saying this is easy: the problem is actually doing it. In principle, the most direct way would be to make DeWitt's approach into rigorous mathematical physics without reference to a fixed background metric. But attempting to do this led immediately into a quagmire which for two decades seemed impassable. It was not until the late 1980s, when Ashtekar [1] found a clever change of variables, that a way around began to seem possible.

We postpone a detailed discussion of these 'new variables' to Section 3. For now, let us simply note that in terms of them, the configuration space of general relativity space is not the space of *metrics* on S, but the space of *connections* on some $\mathrm{SL}(2,\mathbb{C})$ bundle over S. Since $\mathrm{SL}(2,\mathbb{C})$ is a complex Lie group, this configuration space is a complex manifold. It turns out that in the quantum theory, kinematical states should be holomorphic functions on this configuration space. However, using the fact that $\mathrm{SL}(2,\mathbb{C})$ has $\mathrm{SU}(2)$ as a real form, one expects the kinematical state space to be isomorphic to $L^2(\mathcal{A})$, where $\mathcal{A}$ is the space of connections on a certain $\mathrm{SU}(2)$ bundle over S. Thus, at least naively, one expects the physical states of quantum gravity in the new variables formalism to be wavefunctions $\psi \in L^2(\mathcal{A})$ annihilated by certain constraints. These constraints include Hamiltonian and diffeomorphism constraints as before, which express the invariance of ψ under diffeomorphisms of spacetime, but also a 'Gauss law' constraint which expresses the invariance of ψ under gauge transformations.

Now the configuration space of the Yang-Mills equation is also a space of connections, and the only constraint in Yang-Mills theory is the Gauss law. Thus in terms of the new variables, canonical quantum gravity is very similar to $\mathrm{SU}(2)$ Yang-Mills theory with some extra constraints! This is one of the main advantages of the new variables: they allow techniques from Yang-Mills theory to be imported to quantum gravity. Another more technical advantage is that in the new variables, the constraints are polynomial functions of the analogs of position and momentum (and their derivatives). Because the nonrenormalizability of quantum gravity in the traditional metric formulation was closely related to the non-polynomial nature of the constraints, this created a lot of excitement. Unfortunately, at the present time the Hamiltonian constraint still presents thorny problems.

Shortly after the discovery of the new variables, Rovelli and Smolin [45] used them to develop a 'loop representation' of quantum gravity, and this is when the relationship to knot theory became very apparent. The idea of the loop representation of a gauge theory had been developed by Gambini and Trias [31], and basically it consists of writing all the equations one can in terms of 'Wilson loops'. Wilson loops are certain functions on the space of connections: at a point $A \in \mathcal{A}$, the value of the Wilson loop is just the trace of the holonomy of A around some loop α in S, taken in some (finite-dimensional) representation ρ of G, written:

$$\mathrm{tr}(\rho(T\exp\int_\alpha A)).$$

One reason why physicists like them is that they are invariant under gauge transformations, and one expects the physically observable aspects of a gauge theory to be gauge-invariant.

Rovelli and Smolin argued that a state ψ of quantum gravity should give rise to link invariant $\hat{\psi}$, the 'loop transform' of ψ, whose value on the link with components given by the loops $\alpha_1, \ldots, \alpha_n$ is

$$\hat{\psi}(\alpha_1, \ldots, \alpha_n) = \int_{\mathcal{A}} \prod_{i=1}^{n} \mathrm{tr}(\rho((T\exp\int_{\alpha_i} A))\, \psi(A)\, \mathcal{D}A,$$

where ρ is the fundamental representation and $\mathcal{D}A$ is the purely formal 'Lebesgue measure' on $\mathcal{A}$. The reason $\hat{\psi}$ should be a link invariant is that the diffeomorphism constraint says ψ is invariant under diffeomorphisms of S! Of course, this argument is merely heuristic, owing to the mysterious nature of $\mathcal{D}A$, but it is no worse than much of the reasoning in quantum gravity. Indeed, a similar sort of argument led Witten [48] to discover the relation between knot theory and another quantum field theory, Chern-Simons theory. In fact, the link invariant coming from SU(2) Chern-Simons theory, namely the Kauffman bracket, appears to be the loop transform of a state of quantum gravity 'with cosmological constant', meaning that the Einstein equation has been modified to give the vacuum a nonzero stress-energy tensor.

We will not go into this here, since there are a number of expository treatments already [11, 12, 14, 18, 41], but certainly it is one of the main reasons for interest in the interface between knots and quantum gravity.

There is much here that needs to be made more precise in the following sections. However, we are at least in a position now to describe what needs to be done. First, we need to develop integration theory on $\mathcal{A}$, in order to escape the use of purely formal entities like the 'Lebesgue measure' $\mathcal{D}A$. In Section 2 we do this when $\mathcal{A}$ is the space of smooth connections on any principal G-bundle P over a manifold S, where G is a compact connected Lie group and S is real-analytic. There is an especially nice 'generalized measure' on $\mathcal{A}$ which is a substitute for the nonexistent Lebesgue measure. Using it we can define $L^2(\mathcal{A})$, and it turns out that the gauge-invariant subspace of $L^2(\mathcal{A})$ is spanned by 'spin network states'. These are described by graphs analytically embedded in S, with oriented edges labelled by representations of G, and with vertices labelled by intertwining operators from the tensor product of representations labelling 'incoming' edges to the tensor product of representations labelling 'outgoing' edges. Equipped with this mathematical technology, we then return to quantum gravity.

2. Spin Networks

In quantum field theory computations, physicists often try to do integrals over a space $\mathcal{A}$ of connections using a strange thing they call 'Lebesgue measure' on $\mathcal{A}$, usually written $\mathcal{D}A$. This exploits the fact that $\mathcal{A}$ is an affine space, or, arbitrarily choosing one point as the origin, a vector space. The idea is that one should be able to pick a basis for $\mathcal{A}$, thus setting up an isomorphism

$$\mathcal{A} \cong \mathbb{R}^\infty,$$

and then, working in the coordinates defined by this basis, let

$$\mathcal{D}A = \prod_{i=1}^{\infty} dx_i.$$

There are many problems with this idea, however! First, while $\mathcal{A}$ can indeed be identified with an infinite-dimensional vector space, a basis of $\mathcal{A}$ does not really give an isomorphism between $\mathcal{A}$ and an infinite product of copies of $\mathbb{R}$. Second, there is no good theory of infinite products of arbitrary measure spaces. Of course one can be more sophisticated about these issues, but the fact remains that there is no such thing as 'Lebesgue measure' on an infinite-dimensional vector space. Thus it is not surprising that when physicists actually do integrals over $\mathcal{A}$ using 'Lebesgue measure' they often get infinite or ill-defined answers until they perform various sneaky tricks.

On the other hand, there is a different way of doing these integrals which is used in lattice gauge theory. Lattice gauge theory is an approximation to gauge theory on $\mathbb{R}^n$ in which one replaces the continuum by an infinite graph having the points of a lattice as vertices and the line segments between neighboring vertices as edges. A 'connection' on the lattice is simply an assigment of an element g_e of the gauge group G to each edge e of the graph, representing the effect of parallel transport along e. The space of connections in this context is thus very different: it is

$$\mathcal{A} \cong G^\infty,$$

not an infinite product of copies of $\mathbb{R}$, but an infinite product of copies of G! Now, while an infinite product of arbitrary measure spaces is ill-defined, an infinite product of *probability* measure spaces *is* well-defined. (Recall that a probability measure space is a space X with positive measure μ such that $\int_X \mu = 1$.) If G is compact, it is equipped with a very natural probability measure, namely Haar measure, the unique left- and right-invariant Borel measure dg with $\int_G dg = 1$. So in lattice gauge theory with compact gauge group, one can work with the measure

$$\mathcal{D}A = \prod_e dg_e$$

on $\mathcal{A}$, instead of the nonexistent 'Lebesgue measure'. This is precisely what is done.

Now, while lattice gauge theory is easier to make rigorous, it has the disadvantage of being only an approximation to the continuum theories we are really interested in. Indeed, in practice much work in lattice gauge theory is computational in nature. Here one uses not a lattice but a finite graph, so that $\mathcal{A}$ becomes a *finite* product of copies of G, and one numerically calculates integrals over $\mathcal{A}$ by Monte Carlo methods. To apply these results to gauge theory on $\mathbb{R}^n$ one must then investigate not only the 'continuum limit' in which the lattice spacing goes to zero, but also the 'large-volume limit'.

Approximating the continuum by a lattice or a fixed finite graph is particularly distressing in the case of general relativity, which is so intimately connected to the differential geometry of manifolds. It would be nice if one could have ones cake and eat it too, working with connections in the continuum context and preserving diffeomorphism-invariance, but still thinking of the space of connections as something like a product of copies of G. This is precisely what the theory of 'generalized measures' on the space of connections seeks to achieve.

The traditional approach to measure theory is to pick a σ-algebra of 'measurable' subsets of some space X, assign measures to them in a manner satisfying some axioms, and then define a vector space of 'integrable' complex-valued functions on X. Then one figures out how to integrate functions in this space, shows that integration is linear, and shows how to pass limits through integrals under certain conditions. Then, typically, one forgets the proofs of these results and simply uses them. A more modern approach is to shortcut this process and simply choose a space $\mathrm{Fun}(X)$ of functions on X, equip it with a topology, and define a 'generalized measure' μ on X to be a continuous linear functional

$$\mu\colon \mathrm{Fun}(\mathcal{A}) \to \mathbb{C},$$

writing $\mu(f)$ as $\int_X f\, d\mu$ solely out of deference to tradition. Actually, of course, the modern approach complements the old approach rather than replaces it; they are related by many useful theorems, some of which we have summarized elsewhere [9].

Now suppose that P is a principal G-bundle over S, where G is compact and connected, and let $\mathcal{A}$ be the space of smooth connections on P. To define generalized measures on $\mathcal{A}$ we need to choose the space $\mathrm{Fun}(\mathcal{A})$ of functions we want to integrate. Following the idea of the loop representation, for example, we could choose $\mathrm{Fun}(\mathcal{A})$ to be the algebra of functions on $\mathcal{A}$ generated by all Wilson loops, or a completion of this algebra in some topology. In fact, this is how Ashtekar and Isham [5] proceeded in their original attempt to use ideas from the loop representation to set up a rigorous integration theory on $\mathcal{A}$. Subsequent work by Ashtekar, Lewandowski, and the author [6, 7, 9, 10, 37] improved things in a number of ways. For one, it turns out to be simpler to let $\mathrm{Fun}(\mathcal{A})$ contain arbitrary continuous functions of the holonomy of A along *paths* in S. Also, working with smooth paths turns out to be a nuisance, because they can intersect each other in horribly complicated ways. For this reason we shall follow Ashtekar and Lewandowski and require S to be a real-analytic manifold, and work with piecewise real-analytic paths. In a sense this is not so drastic, because every compact smooth manifold can be given a real-analytic structure. One pays a price, however, since one is no longer doing differential geometry in the category of smooth manifolds. Eventually it would be nice to understand the smooth case too.

In any event, let us define $\mathrm{Fun}(A)$ as follows. If $\gamma\colon [0,1] \to S$ is a piecewise analytic path, let $\mathcal{A}_\gamma$ denote the space of smooth maps

$F\colon P_{\gamma(0)} \to P_{\gamma(1)}$ that are compatible with the right action of G on P:

$$F(xg) = F(x)g.$$

Note that for any connection $A \in \mathcal{A}$, the parallel transport map

$$T\exp\int_\gamma A : P_{\gamma(0)} \to P_{\gamma(1)}$$

lies in $\mathcal{A}_\gamma$. If we fix a trivialization of P at the endpoints of γ, we can identify $\mathcal{A}_\gamma$ with the group G. This makes $\mathcal{A}_\gamma$ into a compact manifold in a manner which one can check is independent of the trivialization. Let $\mathrm{Fun}_0(\mathcal{A})$ be the algebra of functions on $\mathcal{A}$ generated by those of the form

$$\psi(A) = f(T\exp\int_{\gamma_1} A, \ldots, T\exp\int_{\gamma_n} A) \tag{1}$$

where $\gamma_1, \ldots, \gamma_n$ are piecewise analytic paths in S, and f is a continuous function on $\mathcal{A}_{\gamma_1} \times \cdots \times \mathcal{A}_{\gamma_n}$. Then let $\mathrm{Fun}(\mathcal{A})$ be the completion of $\mathrm{Fun}_0(\mathcal{A})$ in the sup norm:

$$\|\psi\|_\infty = \sup_{A\in\mathcal{A}} |\psi(A)|.$$

It is easy to check that the Wilson loops lie in this algebra; in fact, they lie in $\mathrm{Fun}_0(\mathcal{A})$.

A generalized measure on $\mathcal{A}$ is then defined to be a continuous linear functional on $\mathrm{Fun}(\mathcal{A})$. This may seem rather abstract, so let us see how to get our hands on one. There exists a nice general recipe for constructing *any* generalized measure on $\mathcal{A}$, which we have described elsewhere [10, 12], but for now let us concentrate on a simple and important example: the uniform generalized measure μ_u. This is a kind of replacement for the nonexistent 'Lebesgue measure' on $\mathcal{A}$, and as we shall see, it is closely modelled after the measure $\mathcal{D}A$ in lattice gauge theory.

To define μ_u, first we define it as a linear functional on $\mathrm{Fun}_0(\mathcal{A})$. Because we are working with real-analytic paths, it turns out that any $\psi \in \mathrm{Fun}_0(\mathcal{A})$ can be written in the form given by equation (1) with the paths $\{\gamma_i\}$ forming an *embedded graph* in S. By this we mean that each path $\gamma_i\colon [0,1] \to S$ is one-to-one and restricts to an embedding of $(0,1)$, and the images $\gamma_i[0,1]$ and $\gamma_j[0,1]$ intersect, if at all, only at their endpoints when $i \neq j$. For functions ψ written in this special form we define

$$\int_{\mathcal{A}} \psi \, d\mu_u = \int_{G^n} f(g_1, \ldots, g_n) dg_1 \cdots dg_n.$$

Here we are using trivializations of P at the endpoints of the paths γ_i to identify $\mathcal{A}_{\gamma_1} \times \cdots \times \mathcal{A}_{\gamma_n}$ with G^n, but the right-hand side is independent of the choice of trivializations, because Haar measure is right- and left-invariant. All the real work goes into checking that the right-hand side does not depend on *how* we wrote ψ in this special form involving an embedded graph. Given that, it is easy to check that μ_u is a linear functional on $\mathrm{Fun}_0(\mathcal{A})$. The bound

$$|\int_{\mathcal{A}} \psi \, d\mu_u| \leq \|\psi\|_\infty$$

then holds because Haar measure is a probability measure. Since $\mathrm{Fun}_0(\mathcal{A})$ is dense in $\mathrm{Fun}(\mathcal{A})$, this bound implies that μ_u extends uniquely to a continuous linear functional on all of $\mathrm{Fun}(\mathcal{A})$, which we again call μ_u. This is the uniform generalized measure on the space of connections!

If we examine what we have just done, the relationship to lattice gauge theory should become clear. Instead of working with a fixed graph embedded in S as one does in lattice gauge theory, we have considered *all possible* embedded graphs $\gamma = \{\gamma_i\}$ in S. Each one of these indeed has the topology of a graph with the paths γ_i as edges and the endpoints $\gamma_i(0), \gamma_i(1)$ as vertices. Following the spirit of lattice gauge theory, we can define a finite-dimensional space of 'connections' on γ,

$$\mathcal{A}_\gamma = \mathcal{A}_{\gamma_1} \times \cdots \times \mathcal{A}_{\gamma_n},$$

on which there is a natural measure, namely a product of copies of Haar measure, one copy for each edge of γ. To do the integral of a function that only depends on the holonomies along the edges of γ, we simply use this natural measure. The key fact is that although a function $\psi \in \mathrm{Fun}_0(\mathcal{A})$ can be expressed in terms of many different embedded graphs in this way — after all, if it can be expressed in terms of a graph γ, it can also be expressed in terms of any graph γ' containing γ — the answer we get for the integral of ψ is independent of this choice. This is why we obtain a well-defined linear functional μ_u on $\mathrm{Fun}_0(\mathcal{A})$, which then extends to all of $\mathrm{Fun}(\mathcal{A})$. In short, we are getting a generalized measure on $\mathcal{A}$ as a kind of 'limit' of measures on the spaces of connections on all graphs embedded in S! This can be made perfectly precise using the language of projective limits [7].

We can now define the Hilbert space $L^2(\mathcal{A})$ as the completion of $\mathrm{Fun}(\mathcal{A})$ with respect to the norm

$$\|\psi\|_2 = (\int_{\mathcal{A}} |\psi|^2 \, d\mu_u)^{1/2}.$$

In the special case where $G = \mathrm{SU}(2)$ and S is a compact oriented 3-manifold, this Hilbert space will be our space of 'kinematical states' for quantum gravity in the new variables formalism. The Gauss law constraint then amounts to requiring that the state $\psi \in L^2(\mathcal{A})$ be invariant under gauge transformations, so it is important to find gauge-invariant vectors in $L^2(\mathcal{A})$. Finding these vectors is actually of interest no matter what G and S happen to be. It turns out that these vectors can also be thought of as wavefunctions on the quotient of $\mathcal{A}$ by the group $\mathcal{G}$ of smooth gauge transformations [12], so we will denote the space of gauge-invariant vectors in $L^2(\mathcal{A})$ by $L^2(\mathcal{A}/\mathcal{G})$.

The most obvious examples of vectors in $L^2(\mathcal{A}/\mathcal{G})$ are the Wilson loops. If we have an analytic loop α in S and a representation ρ of G, the function

$$\psi(A) = \mathrm{tr}(\rho(T\exp\int_\alpha A))$$

is gauge-invariant, and it lies in $L^2(\mathcal{A})$. More generally, any product of Wilson loops is a gauge-invariant element of $L^2(\mathcal{A})$.

More generally still, we can get vectors in $L^2(\mathcal{A}/\mathcal{G})$ from 'spin networks' [13]. Take a graph γ embedded in S and label each of its edges e with a representation ρ_e of G. Let

$$H_e(A) = \rho_e(T\exp\int_e A).$$

If we trivialize P at the vertices of γ, and pick a basis for ρ_e, we can think of $H_e(A)$ as a matrix $H_e(A)^i_j$. Now form the tensor product of all these matrices, one for each edge of γ. We get a big tensor $H(A)$ having one superscript and one subscript for each edge; it is too ugly to bother writing down, but we hope the reader gets the idea.

Next, for each vertex v of γ let $S(v)$ be the set of edges having v as 'source' — where the source of an edge γ_i is defined to be $\gamma_i(0)$ — and let $T(v)$ be the set of edges having v as 'target' — where the target of γ_i is $\gamma_i(1)$. For each vertex v, pick an intertwining operator

$$I_v \colon \bigotimes_{e\in T(v)} \rho_e \to \bigotimes_{e\in S(v)} \rho_e.$$

We can think of I_v as a tensor with one superscript for each edge $e \in T(v)$ and one subscript for each edge $e \in S(v)$. Then form the tensor product of all these tensors I_v, one for each vertex. We get a big tensor I. Then form the tensor product of $H(A) \otimes I$. Note that each superscript of $H(A)$ corresponds to a particular subscript of I and vice versa, because each edge of γ lies in $S(v)$ for one vertex v and lies in $T(w)$ for one vertex w. So we can contract the tensor $H(A) \otimes I$ to get a number, which of course depends on A. This is our 'spin network state' $\psi(A)$. Note that a Wilson loop is just a special case of a spin network with only one edge and one vertex, with the intertwining operator taken to be the identity operator.

One can show directly from this explicit definition that the spin network states are gauge-invariant and lie in $L^2(\mathcal{A})$, in fact in $\mathrm{Fun}_0(\mathcal{A})$. But we want to show more: we want to show they span the whole space of gauge-invariant vectors in $L^2(\mathcal{A})$. For this a more abstract approach is better.

The group $\mathcal{G}$ acts on $\mathcal{A}$, and in fact there is a unitary representation of $\mathcal{G}$ on $L^2(\mathcal{A})$ given by

$$g\psi(A) = \psi(g^{-1}A).$$

To see this it is useful to introduce work 'one embedded graph at a time'. One can show that for any embedded graph γ, $L^2(\mathcal{A}_\gamma)$ can be identified with the smallest closed subspace of $L^2(\mathcal{A})$ containing all functions of the form given in equation (1). If we do a gauge transformation $g \in \mathcal{G}$ on the connection A, the holonomy along any edge γ_i transforms to

$$T\exp\int_{\gamma_i} gA = g(\gamma_i(1))\,(T\exp\int_{\gamma_i} A)\,g(\gamma_i(0))^{-1}, \tag{2}$$

so if ψ lies in $L^2(\mathcal{A}_\gamma)$, so does $g\psi$. Thus we get a representation of $\mathcal{G}$ on each subspace $L^2(\mathcal{A}_\gamma)$, and these representations are unitary because Haar measure is left- and right-invariant. Since the union of the subspaces $L^2(\mathcal{A}_\gamma)$ is dense in $L^2(\mathcal{A})$, we obtain a unitary representation of $\mathcal{G}$ on all of $L^2(\mathcal{A})$.

Let $L^2(\mathcal{A}_\gamma/\mathcal{G}_\gamma)$ denote the gauge-invariant subspace of $L^2(\mathcal{A}_\gamma)$. Since the action of $\mathcal{G}$ on $L^2(\mathcal{A})$ preserves each subspace $L^2(\mathcal{A}_\gamma)$, the union of the subspaces $L^2(\mathcal{A}_\gamma/\mathcal{G}_\gamma)$ must span $L^2(\mathcal{A}/\mathcal{G})$. So, what does $L^2(\mathcal{A}_\gamma/\mathcal{G}_\gamma)$ look like? For this we need a very precise picture of the action of $\mathcal{G}$ on $L^2(\mathcal{A}_\gamma)$.

Write E for the set of edges of γ, and V for the set of vertices. Then picking a trivialization of P over the vertices of γ, we obtain an isomorphism

$$\begin{aligned} L^2(\mathcal{A}_\gamma) &\cong L^2(G^E) \\ &\cong \bigotimes_{e\in E} L^2(G). \end{aligned}$$

To see how $\mathcal{G}$ acts on the right-hand side, we use the Peter-Weyl theorem. Note that $G \times G$ acts on G by

$$(g_1, g_2)(h) = g_1 h g_2^{-1},$$

giving a unitary representation of $G \times G$ on $L^2(G)$. The Peter-Weyl theorem describes how $L^2(G)$ decomposes into irreducible unitary representations of $G \times G$. These are all of the form $\rho_1 \otimes \rho_2$, where ρ_1 and ρ_2 are irreducible unitary representations of G. Let R be a set containing one irreducible unitary representation of G from each equivalence class. Then the Peter-Weyl theorem says that

$$L^2(G) \cong \bigoplus_{\rho \in R} \rho \otimes \rho^*.$$

It follows that

$$L^2(\mathcal{A}_\gamma) \cong \bigotimes_{e \in E} \bigoplus_{\rho \in R} \rho \otimes \rho^*,$$

and in terms of this description, any $g \in \mathcal{G}$ acts on $L^2(\mathcal{A}_\gamma)$ as the operator

$$\bigotimes_{e \in E} \bigoplus_{\rho \in R} \rho(g(s(e)) \otimes \rho^*(g(t(e))).$$

The reason is that the holonomy of A along each edge e transforms in precisely the manner suited to applying the Peter-Weyl theorem, as we saw in equation (2).

Next, using the associativity of tensor product over direct sum, we obtain

$$L^2(\mathcal{A}_\gamma) \cong \bigoplus_{\rho \in R^E} \bigotimes_{e \in E} \rho_e \otimes \rho_e^*,$$

where R^E is the set of all labellings of edges $e \in E$ by representations $\rho_e \in R$. Grouping the edges by their source and target, this gives

$$L^2(\mathcal{A}_\gamma) \cong \bigoplus_{\rho \in R^E} \bigotimes_{v \in V} \left(\bigotimes_{e \in S(v)} \rho_e \otimes \bigotimes_{e \in T(v)} \rho_e^* \right).$$

In these terms any $g \in \mathcal{G}$ acts on $L^2(\mathcal{A}_\gamma)$ as the operator

$$\bigoplus_{\rho \in R^E} \bigotimes_{v \in V} \left(\bigotimes_{e \in S(v)} \rho_e(g_v) \otimes \bigotimes_{e \in T(v)} \rho_e^*(g_v) \right).$$

Thus we have

$$L^2(\mathcal{A}_\gamma/\mathcal{G}_\gamma) \cong \bigoplus_{\rho \in R^E} \bigotimes_{v \in V} \mathrm{Inv} \left(\bigotimes_{e \in S(v)} \rho_e \otimes \bigotimes_{e \in T(v)} \rho_e^* \right),$$

where for any representation λ of G, $\mathrm{Inv}(\lambda)$ is the subspace of G-invariant vectors. But $\mathrm{Inv}(\lambda_2 \otimes \lambda_1^*)$ is isomorphic to the space $\mathrm{Hom}(\lambda_1, \lambda_2)$

of intertwining operators from λ_1 to λ_2, so finally we have

$$L^2(\mathcal{A}_\gamma/\mathcal{G}_\gamma) \cong \bigoplus_{\rho\in R^E} \bigotimes_{v\in V} \mathrm{Hom}\left(\bigotimes_{e\in T(v)} \rho_e, \bigotimes_{e\in S(v)} \rho_e\right).$$

In other words, a complete set of gauge-invariant vectors in $L^2(\mathcal{A}_\gamma)$ is obtained by labelling each edge of γ with an irreducible unitary representation of G and labelling each vertex of γ with an intertwining operator from the tensor product of representations labelling incoming edges to the tensor product of representations labelling outgoing edges. If one unravels the logic of the proof, one sees that these are special cases of the spin network states described more explicitly above!

Our work has shown that spin network states span $L^2(\mathcal{A}/\mathcal{G})$. But in fact we have shown more. For each embedded graph γ we obtain an orthonormal basis of $L^2(\mathcal{A}_\gamma/\mathcal{G}_\gamma)$ by letting ρ range over all labellings of edges by irreducible unitary representations, and, for each ρ, picking an orthonormal basis of

$$\mathrm{Hom}\left(\bigotimes_{e\in T(v)} \rho_e, \bigotimes_{e\in S(v)} \rho_e\right)$$

for each vertex v. While it is a nuisance to assemble all these orthonormal bases into a single orthonormal basis for all of $L^2(\mathcal{A}/\mathcal{G})$, for practical computations having an orthonormal basis for each graph is usually sufficient.

Of course, for *really* practical computations one might need a recipe for picking an orthonormal basis of the space of intertwining operators

$$\mathrm{Hom}\left(\bigotimes_{e\in T(v)} \rho_e, \bigotimes_{e\in S(v)} \rho_e\right).$$

For example, Brügmann [19] has tried some computer simulations of quantum gravity using spin networks, and computers are notorious for wanting everything to be very explicit. To pick such an orthonormal basis, one needs to understand the representation theory of G. Luckily, the representation theory of SU(2) is quite simple, so let us consider that case. A more detailed treatment can be found in the work of Rovelli and Smolin [46].

The irreducible unitary representations of SU(2) can be labelled by their dimension $d = 1, 2, 3, \ldots$, or equivalently by their 'spin' $j = 0, \frac{1}{2}, 1, \ldots$, where $d = 2j+1$. The tensor product of two such representations decomposes as follows:

$$j_1 \otimes j_2 = |j_1 - j_2| \;\oplus\; |j_1 - j_2| + 1 \;\oplus\; \cdots \;\oplus\; j_1 + j_2.$$

This means that for a *trivalent* graph γ we do not need to worry much about labelling the vertices with intertwining operators. Suppose, for example, that a vertex has two incoming edges labelled with spins j_1 and j_2, and one outgoing edge labelled j_3. Unless the Clebsch-Gordon condition

$$|j_1 - j_2| \leq j_3 \leq j_1 + j_2, \qquad j_1 + j_2 + j_3 \in \mathbb{Z}$$

holds, the representation j_3 will not appear as a summand of of the tensor product $j_1 \otimes j_2$, and there is no way to get a nonzero spin network state from this labelling. If the Clebsch-Gordon condition does hold, j_3 appears with multiplicity 1 in $j_1 \otimes j_2$, so up to a constant factor which one must fix in some standard way, there is no choice about which intertwining operator to pick.

Graphs that are not trivalent can be reduced to the trivalent case as follows. Take each n-valent vertex and replace it in ones mind by a n-leaved tree each of whose vertices is trivalent. If the original vertex was n-valent, this tree will have $n - 3$ new 'internal edges'. A basis of intertwining operators for the original vertex is then given by all labellings of these internal edges by spins satisfying the Clebsch-Gordon condition. Of course, there are many different trees with n leaves, so there is a great deal of arbitrariness in this choice of basis. However, to change from one basis to another simply requires repeated use of some matrices known as the $6j$ symbols, which are familiar in SU(2) representation theory and whose generalizations play an important role in topological quantum field theory [26, 36].

Before returning to a more thorough treatment of the new variables and the use of spin networks of quantum gravity, let us say a little about the history of spin networks. Because of the immense difficulties they have had in quantizing gravity, physicists have often considered the possibility that our picture of spacetime as a manifold breaks down at the Planck scale. This has led to various attempts to reformulate physics in terms of more discrete, or combinatorial, ideas. In fact, spin networks were invented in the early 1970s by Penrose [39] in one such attempt. However, while our spin networks involve graphs *embedded* in a pre-existing manifold that represents space, his spin networks were *abstract* graphs labelled by representations and intertwining operators. In other words, rather than serving as a tool for describing the geometry of a spacetime manifold, his spin networks were intended as a purely combinatorial substitute for a spacetime manifold. The problem with this sort of radical idea has always been bridging the gap between it and existing theories of physics. Thus the more recent discovery that spin networks also arise naturally in attempts to quantize Einstein's equation is quite intriguing.

If we succeeded in constructing quantum gravity as a field theory on a manifold, we might simply decide to forget about more combinatorial approaches to quantum gravity. However, as we shall see, the Hamiltonian constraint poses serious problems even in the new variables formalism, so-called 'ultraviolet problems', which might be due to falsely extrapolating our picture of spacetime as a manifold to arbitrarily small length scales. Thus it is still worth keeping in mind the possibility that some combinatorial approach is the fundamental one. For this reason, any clues about the relation between 'abstract' and 'embedded' spin networks are likely to be interesting.

Luckily, there are quite a few clues along these lines! While Penrose's ideas yield a purely combinatorial recipe for defining an invariant of abstract SU(2) spin networks, this invariant can also be described in terms of spin networks embedded in $\mathbb{R}^3$. One can obtain a generalized measure μ on the space $\mathcal{A}$ of SU(2) connections on $\mathbb{R}^3$ by taking a finite Borel measure μ_0 on the subspace $\mathcal{A}_0$ of flat connections, and defining

$$\int_{\mathcal{A}} \psi(A) d\mu = \int_{\mathcal{A}_0} \psi(A) d\mu_0. \tag{3}$$

Then if ψ is a spin network state associated to some graph γ, the integral $\int_{\mathcal{A}} \psi(A) d\mu$ does not depend on the choice of embedding of γ, essentially because the holonomy of a *flat* connection along a path does not change when the path is deformed. Thus we obtain an invariant of abstract spin networks. If μ_0 is a probability measure, this is the same invariant that one can define combinatorially following the ideas of Penrose (while carefully adjusting his sign conventions [46]).

A striking generalization of this idea arises from the work of Witten [48] on Chern-Simons theory and the Jones polynomial, and its subsequent reinterpretation using quantum groups by Reshtekhin and Turaev [42]. Heuristically, Chern-Simons theory defines an isotopy invariant of spin networks embedded in S^3 by means of a measure on the space of connections:

$$\int_{\mathcal{A}} \psi(A) e^{\frac{ik}{4\pi} \int_{S^3} \mathrm{tr}(A \wedge dA + \frac{2}{3} A \wedge A \wedge A)} \mathcal{D}A \tag{4}$$

Indeed, Witten was able by formal manipulations to compute these invariants in certain cases. Unfortunately, unlike equation (3), this formula is difficult to make rigorous. It involves the purely formal 'Lebesgue measure' $\mathcal{D}A$, and replacing $\mathcal{D}A$ by the uniform generalized measure does not help much, since the quantity integrated against it does not lie in Fun($\mathcal{A}$). In fact, one cannot really compute these spin network invariants using a generalized measure in our precise sense of the term [12]. One reason, though not the only one, is that the invariants depend on a choice of extra structure known as a 'framing'. Nonetheless, Reshetikhin and Turaev were able to rigorously construct these isotopy invariants of framed embedded spin networks by a combinatorial procedure that generalizes Penrose's recipe to quantum groups. In the case of links labelled by the spin-$\frac{1}{2}$ representation of SU(2), this invariant is just the Kauffman bracket.

These ideas interact with the subject of quantum gravity in many ways. In Section 1 we cited some expository accounts of the relation between Chern-Simons theory and quantum gravity in 4-dimensional spacetime. However, spin networks and Chern-Simons theory are also closely related to quantum gravity with cosmological constant in 3-dimensional spacetime [28, 36, 40, 43, 44, 47, 49]. In this case, quantum gravity can be rigorously described *either* by starting with Einstein's equation on a smooth 3-manifold, phrased in terms of an appropriate version of the new variables, *or* purely combinatorially in terms of a simplicial 3-manifold with edges labelled by spins. Better yet, although there is still a bit of work left in proving this rigorously, it appears that the two approaches are equivalent! In short, in this toy model the question as to whether spacetime is continuous or discrete has no simple yes-or-no answer; in some sense, the answer is 'both'.

It is natural to wonder if the same might be true in 4 dimensions. There are a few hints here. One the one hand, Einstein's equation in the new variables formulation has a simpler relative called BF theory [12], which is much easier to canonically quantize. On the other hand, Crane and Yetter have combinatorially constructed a relatively simple topological quantum field theory in 4 dimensions using spin networks [22]. It is not known if these are equivalent, though there are some clues suggesting it, such as the relation of both to Chern-Simons theory. This would be worth understanding better, even though we expect full-fledged quantum gravity in 4 dimensions to be more complicated.

3. The New Variables

Now let us return to general relativity and describe more precisely how the new variables make it look more like other gauge theories. In its original formulation, general relativity is all about a *metric* on spacetime, while gauge theories are all about a *connection* on some bundle over spacetime. Of course there is a connection involved in general relativity too, the Levi-Civita connection, but this is traditionally regarded as a subsidiary entity, since it can be computed starting from the metric. To emphasize the gauge-theoretic aspects of general relativity, one needs to rewrite it so a connection has the starring role, and the metric appears as more of a minor character. There have been many different attempts to do this, and for a more thorough exploration of them the reader will have to turn elsewhere [38]. Here we only describe two: the Palatini formalism, and the Plebanski formalism. The latter is the one directly related to the 'new variables', but the former serves as a good warmup exercise. This section makes somewhat greater demands on the reader's acquaintance with differential geometry, and also uses some variational calculus. Luckily, there happens to be a textbook that explains most of what we need [14]. We will skip over all sorts of important subtleties, which are discussed in Ashtekar's books [2, 3].

In the Palatini formalism there will be two basic fields, a connection and a 'soldering form'. The clever idea (in the modern version of this approach) is to fix an oriented bundle $\mathcal{T}$ over M that is isomorphic to the tangent bundle TM, but not canonically so. We can think of $\mathcal{T}$ as a kind of 'imitation tangent bundle'. Physicists usually call it, or any of its fibers, the 'internal space'. We assume $\mathcal{T}$ is equipped with a Lorentzian metric η — the 'internal metric'— and assume that the spacetime metric g is obtained from η via an isomorphism $e\colon TM \to \mathcal{T}$. In other words,

$$g(v, w) = \eta(e(v), e(w)) \tag{5}$$

for any vector fields v and w. We may regard e as a $\mathcal{T}$-valued 1-form, and then it is called the soldering form. We should add, in the interests of cultural literacy, that e is also called the 'coframe field', while e^{-1} is known as the 'frame field', 'tetrad field', or 'vierbein'.

In Palatini formalism the two basic fields are this soldering form e and a connection A on $\mathcal{T}$ preserving the metric η, usually called a 'Lorentz connection'. One virtue of this formalism is that it makes sense even when $e\colon TM \to \mathcal{T}$ is *not* an isomorphism. Thus the Palatini formalism extends general relativity to certain cases where the metric g is degenerate.

In computations it is handy to use differential forms on M taking values in the exterior algebra bundle $\Lambda\mathcal{T}$. These form an algebra, where the product (written as $\wedge$) is built from the exterior product in $\Lambda\mathcal{T}$ together with the usual wedge product of differential forms. In complete analogy with the volume form on an oriented Lorentzian manifold, the orientation and internal metric on $\mathcal{T}$ give rise to an 'internal volume form', i.e. a section ν of $\Lambda^4\mathcal{T}$. This in turn yields a map from $\Lambda^4\mathcal{T}$-valued forms to ordinary differential forms, denoted by 'tr', and given by

$$\mathrm{tr}(\nu \otimes \omega) = \omega$$

for any differential form ω on M. Now the curvature of the connection A can, as usual, be regarded as an $\mathrm{End}(\mathcal{T})$-valued 2-form, but the internal metric provides an isomorphism $\mathcal{T} \cong \mathcal{T}^*$, so we may think of it as $\mathcal{T} \otimes \mathcal{T}$-valued, and then the fact that A is metric-preserving means the curvature actually takes values in $\Lambda^2 \mathcal{T}$. We call this $\Lambda^2 \mathcal{T}$-valued 2-form F.

One can then obtain the vacuum Einstein equation from a variational principle, starting with the 'Palatini action' given by

$$S_{Pal}(A, e) = \int_M \mathrm{tr}(e \wedge e \wedge F).$$

Let us sketch how this goes. The idea is to compute the variation δS_{Pal}, demand that it vanish for all compactly supported variations δA and δe, and see what this implies. We will need to use the wonderful formula

$$\delta F = d_A \delta A, \tag{6}$$

where d_A denotes the exterior covariant derivative of $\Lambda \mathcal{T}$-valued forms with respect to the connection A, and δA is treated as a $\Lambda^2 \mathcal{T}$-valued 1-form. We begin by computing the variation, or differential, of S_{Pal} as a function of A and e:

$$\begin{aligned} \delta S_{Pal} &= \int \delta \mathrm{tr}(e \wedge e \wedge F) \\ &= \int \mathrm{tr}(2\delta e \wedge e \wedge F + e \wedge e \wedge \delta F), \end{aligned}$$

using the standard rules for differentiation. Using equation (6) we obtain

$$\delta S_{Pal} = \int \mathrm{tr}(2\delta e \wedge e \wedge F + e \wedge e \wedge d_A \delta A).$$

Finally, integrating by parts as one always does in these variational computations,

$$\delta S_{Pal} = 2 \int \mathrm{tr}(\delta e \wedge e \wedge F - e \wedge d_A e \wedge \delta A).$$

This only vanishes for *all* compactly supported δA, δe if

$$e \wedge F = 0, \qquad e \wedge d_A e = 0.$$

These are our 'classical equations of motion'.

When e is an isomorphism, these equations are really just the vacuum Einstein equation in disguise. First, one can show that in this case $e \wedge d_A e = 0$ implies $d_A e = 0$. Then, using the isomorphism $e \colon TM \to \mathcal{T}$ to transfer A to a metric-preserving connection on TM, say Γ, the equation $d_A e = 0$ implies that Γ is torsion-free, hence equal to the Levi-Civita connection of g. This lets us translate the equation $e \wedge F = 0$ into an equation about the curvature of Γ, i.e., the Riemann tensor. When we do so, we obtain the vacuum Einstein equation!

The Plebanski formalism works in much the same way. Indeed, at first it may seem like just an unnecessarily complicated version of the Palatini formalism. As noted by Ashtekar, however, the *constraints* work out much more nicely when one tries to canonically quantize the theory. The Plebanski formalism is also called the 'self-dual' formalism, because it takes advantage of self-duality, a very special feature of 4 dimensions. Recall that given a 4-dimensional real inner product space V, the Hodge star operator maps the second exterior power of V to itself:

$$\star\colon \Lambda^2 V \to \Lambda^2 V.$$

Since $\star^2 = 1$, we can split any element $\omega \in \Lambda^2 V$ into a self-dual and an anti-self-dual part:

$$\omega = \omega_+ + \omega_-, \qquad \star\omega_\pm = \pm\omega_\pm.$$

Another way to think of this is as follows. The inner product gives an isomorphism between $\mathrm{End}(V) = V \otimes V^*$ and $V \otimes V$, which restricts to an isomorphism between $\mathfrak{so}(V)$ and $\Lambda^2 V$. The splitting of $\Lambda^2 V$ into self-dual and anti-self-dual parts then turns out to correspond to a splitting of the 6-dimensional Lie algebra $\mathfrak{so}(V)$ into two 3-dimensional ones. Taking $V = \mathbb{R}^4$ this gives

$$\mathfrak{so}(4) \cong \mathfrak{so}(3) \oplus \mathfrak{so}(3). \tag{7}$$

These simple facts have all sorts of ramifications for 4-dimensional physics and topology, especially in gauge theory, where a connection on a Riemannian manifold having self-dual curvature 2-form is automatically a solution of the Yang-Mills equation, called an 'instanton' [29]. When M is compact, the space of instantons modulo gauge transformations is finite-dimensional, and one can extract powerful invariants of the smooth structure of M from this 'moduli space' using techniques developed by Donaldson and others [25]. For example, one can use these to show that $\mathbb{R}^4$ admits uncountably many different smooth structures, unlike any other $\mathbb{R}^n$! Seiberg and Witten have recently simplified some aspects of Donaldson theory by further recourse to ideas from physics, but self-duality still plays a key role.

The sort of self-duality relevant to the Plebanski formalism is rather different, though with tantalizing relationships to the above [21]. First, given a Lorentzian rather than Riemannian 4-manifold, one has $\star^2 =$

-1 on 2-forms. This means that the eigenvalues of the Hodge star operator are $\pm i$, so one needs to work with *complex-valued* 2-forms to take advantage of self-duality. When we tensor both sides of equation (7) with $\mathbb{C}$, we obtain

$$\mathfrak{so}(4,\mathbb{C}) \cong \mathfrak{sl}(2,\mathbb{C}) \oplus \mathfrak{sl}(2,\mathbb{C}).$$

For these reasons, the Plebanski formalism applies most naturally to *complexified* general relativity, and some extra work is needed to restrict to real-valued metrics.

Second, in the Plebanski formalism self-duality first shows up not with respect to 'honest' 2-forms, but with respect to sections of $\Lambda^2\mathbb{C}\mathcal{T}$, the second exterior power of the complexified internal space. The internal metric η extends by complexification to a bilinear pairing on $\mathbb{C}\mathcal{T}$, which we again call η, and this together with the orientation give rise to an 'internal' Hodge star operator $*$ acting on sections of $\Lambda^2\mathbb{C}\mathcal{T}$. We can thus split sections of this bundle into self-dual and anti-self-dual parts:

$$\omega = \omega_+ + \omega_-, \qquad *\omega_\pm = \pm i\omega_\pm.$$

Now, the orthonormal frame bundle of $\mathbb{C}\mathcal{T}$ is a principal bundle P with structure group $\mathrm{SO}(4,\mathbb{C})$. Assume we have a spin structure for $\mathbb{C}\mathcal{T}$, that is, a double cover $\tilde{P}$ of P with structure group $\widetilde{SO}(4,\mathbb{C}) = \mathrm{SL}(2,\mathbb{C}) \oplus \mathrm{SL}(2,\mathbb{C})$. Then $\tilde{P}$ is the sum $P_+ \oplus P_-$ of two principal bundles with structure group $\mathrm{SL}(2,\mathbb{C})$. A metric-preserving connection on $\mathbb{C}\mathcal{T}$ is usually called a 'complex Lorentz connection'. By the above, a complex Lorentz connection A is equivalent to a pair of connections $A_\pm$ on $P_\pm$, which we call the self-dual and anti-self-dual parts of A. The curvature of $A_\pm$ is a 2-form $F_\pm$ taking values in the associated bundle $\mathrm{Ad}P_\pm$, but sections of these bundles can be identified with self-dual (resp. anti-self-dual) sections of $\Lambda^2\mathbb{C}\mathcal{T}$. The curvature of A is a $\Lambda^2\mathbb{C}\mathcal{T}$-valued 2-form F with $F = F_+ + F_-$.

The two basic fields in the Plebanski formalism are a *complex-valued soldering form*, that is, 1-form on M with values in $\mathbb{C}\mathcal{T}$, and a *self-dual Lorentz connection* A_+, that is, a connection on P_+. The Plebanski action is given by:

$$S_{Ple}(A_+, e) = \int_M \mathrm{tr}(e \wedge e \wedge F_+)$$

This is very much like the Palatini action, and, much as before, the classical equations of motion are

$$e \wedge F_+ = 0, \qquad e \wedge d_{A_+} e = 0.$$

These give Einstein's equation for complex-valued metrics on M when e defines an isomorphism between the complexified tangent bundle of M and $\mathbb{C}\mathcal{T}$. To see this, first note that e gives rise to a complex-valued metric g on M by equation (5). Moreover, we can use e to transfer A_+ to a connection Γ_+ on the complexified tangent bundle of M. The equation $e \wedge d_{A_+} e = 0$ then implies Γ_+ is torsion-free, hence equal to the self-dual part of the Levi-Civita connection Γ of g. We can transfer Γ back to $\mathbb{C}\mathcal{T}$ and get a connection A having A_+ as its self-dual part. Then it turns out that $e \wedge F_+ = e \wedge F$, so as in the Palatini formalism we obtain Einstein's equation for g.

In terms of these 'new variables', as they are often called, the configuration space of general relativity is no longer the space of all metrics on S. Instead, it is the space $\mathcal{A}_+$ of all connections on P_+ restricted to S, or more precisely, to some fixed spacelike slice $S_t \subset M$. As in the metric formulation, one can separate the equations of motion into evolutionary equations and constraints. We shall not go through the calculations here, but simply state some of the results [2, 3]. Recall that in the metric formulation of general relativity, the analogs of the position and momentum are the fields q_{ij} and p^{ij} on S. In the new variables, the analogs of position and momentum are instead the connection A_+ and the field $(e \wedge e)_+$ restricted to S. Since P_+ is trivial when restricted to S, A_+ can be identified with an $\mathfrak{sl}(2,\mathbb{C})$-valued 1-form on S, usually denoted simply by A. Similarly, $(e \wedge e)_+$ can be identified with an $\mathfrak{sl}(2,\mathbb{C})$-valued 2-form on S, but physicists often use the isomorphism

$$\Lambda^2 T^*S \cong TS \otimes \Lambda^3 T^*S,$$

given by the interior product, to treat it as a 'densitized vector field' on S with values in $\mathfrak{sl}(2,\mathbb{C})$, usually denoted by $\tilde{E}$. Actually, using coordinates, they often write A and $\tilde{E}$ as A_i^a and $\tilde{E}^{ia}$, where $a = 1,2,3$ indexes a basis of $\mathfrak{sl}(2,\mathbb{C})$ that is orthonormal with respect to the Killing form.

In these terms, the constraints in the new variables approach are

$$\mathcal{G}^a = \partial_i \tilde{E}^{ia} + [A_i, \tilde{E}^i]^a, \tag{8}$$

$$\mathcal{H}_j = F_{ij}^a \tilde{E}_a^i, \tag{9}$$

and

$$\mathcal{H} = \epsilon_{abc} F_{ij}^c \tilde{E}^{ia} \tilde{E}^{jb}, \tag{10}$$

where F is the curvature of A and ϵ_{abc} is the completely antisymmetric tensor with $\epsilon_{123} = 1$.

It is probably not obvious from our description where the advantage of the self-dual formalism over the Palatini formalism lies. The key point is the construction of the field $\tilde{E}$ from $(e \wedge e)_+$. One can similarly construct $\tilde{E}$ from $e \wedge e$ in the Palatini formalism, but then $\tilde{E}$ winds up being subject to extra constraints [3] which negate the advantages of this approach. In the self-dual formalism, no conditions on $\tilde{E}$ need hold for it to come from a complex soldering form e.

4. Canonical Quantization

Now let us try to canonically quantize gravity in terms of the new variables, with an eye to the importance of spin networks. The basic idea of DeWitt's approach goes over to this context with only a few small modifications. In its naive form, the idea would be to define a Hilbert space $L^2(\mathcal{A}_+)$ and define operators on this space by

$$(\hat{A}_i^a \psi)(A) = A_i^a \psi(A), \quad (\hat{\tilde{E}}_a^i \psi)(A) = \hbar \frac{\delta \psi}{\delta A_i^a}(A).$$

(Again, we have suppressed the dependence of these operators on the point in our spacelike slice.) We would then substitute these operators for A and $\tilde{E}$ in formulas (8-10), obtaining quantized constraints $\hat{\mathcal{H}}$, $\hat{\mathcal{H}}_j$, and $\hat{\mathcal{G}}^a$. Physical states should then be annihilated by these constraints.

There are, however, a number of subtleties that we need to address in order to do things right. First there is the fact that the Plebanski formalism most naturally describes complex general relativity, and needs some adjustment to become a theory of honest real-valued Lorentzian metrics. This issue has been a murky one for some time, and only now is it beginning to become clear. At the classical level one can simply impose extra constraints saying that the metric is Lorentzian. This would not be very nice in the quantum theory, though, because the whole point of the new variables was to simplify the constraints! Luckily it appears that at the quantum level we can deal with the issue in quite a different way, namely by restricting our attention to a space $HL^2(\mathcal{A}_+)$ of holomorphic wavefunctions on $\mathcal{A}_+$. Rather than really explaining this here, let us merely note that it is closely related to the Bargmann-Segal formulation of the quantum theory of a particle on a line [3, 4]. The Bargmann-Segal formulation makes use of an isomorphism between $L^2(\mathbb{R})$ and the Hilbert space $HL^2(\mathbb{C})$ consisting of holomorphic functions that are square-integrable with respect to a Gaussian measure. A similar theory for complex Lie groups and their compact real forms, due to Hall [32], gives an isomorphism between a Hilbert space $HL^2(\mathrm{SL}(2,\mathbb{C}))$ and the Hilbert space $L^2(\mathrm{SU}(2))$ defined using Haar measure. A generalization of this [8] gives an isomorphism, the 'coherent state transform', between a Hilbert space $HL^2(\mathcal{A}_+)$ and the space $L^2(\mathcal{A})$, where $\mathcal{A}$ is the space of connections on an SU(2) bundle over S. Thus the theory described in Section 2 is relevant to quantum gravity even though $\mathrm{SL}(2,\mathbb{C})$ is not compact.

Second, there is the issue of rigorously interpreting the constraints and finding solutions to them. When we try to replace the fields A and $\tilde{E}$ by $\hat{A}$ and $\hat{\tilde{E}}$ in formulas (8-10), we run into difficulties. One problem is that operators do not commute, so different orderings of the same polynomial in the classical fields can have different meanings at the quantum level. The orderings given here seem best when one wants operators on some space of functions on $\mathcal{A}_+$; elsewhere one may see the opposite orderings [12], but that is the natural consequence of working dually on some space of generalized measures. A more profound problem is that $\hat{A}$ and $\hat{\tilde{E}}$ are not really operator-valued functions, but only operator-valued distributions — one must integrate them against 'test functions' (really bundle sections) to obtain operators — and like ordinary distributions, the product of operator-valued distributions is only defined under special conditions, or using special tricks.

Luckily, the Gauss law constraint and diffeomorphism constraint have simple geometrical interpretations which relieve us of the need for making sense of the quantum analogs of formulas (8) and (9). At the classical level, functions of the form $\int_S X_a \mathcal{G}^a$ generate all the flows on the phase space $T^*\mathcal{A}$ that correspond to the action of one-parameter groups of gauge transformations on $T^*\mathcal{A}$. So at the quantum level we can interpret the Gauss law constraint as saying that a state is invariant under 'small' gauge transformations, that is, those lying in the component of $\mathcal{G}$ containing the identity. One can show that any state $\psi \in L^2(\mathcal{A})$ that is invariant under small gauge transformations is invariant under all gauge transformations. As we have seen, there is an ample supply of such states, and the space of these, which we write as $L^2(\mathcal{A}/\mathcal{G})$, is spanned by spin network states.

Similarly, any function on phase space of the form $\int_S N^i \mathcal{H}_i$ generates a flow corresponding to the action of a one-parameter group of diffeomorphisms. Thus at the quantum level we can interpret the diffeomorphism constraint as saying that a state is invariant under the group $\mathrm{Diff}_0(S)$ of small analytic diffeomorphisms, that is, those lying in the connected component containing the identity. Here we run into a problem. It is easy to see that the only element of $\mathrm{Fun}_0(\mathcal{A})$ invariant under $\mathrm{Diff}_0(S)$ is the function 1; any other can be written as in equation (1) only for some nonempty graph γ, and then there is a small diffeomorphism taking it to some other function, in fact one orthogonal to it in $L^2(\mathcal{A})$. A simple approximation argument then shows that no $\psi \in L^2(\mathcal{A})$ except $\psi = 1$ is invariant under $\mathrm{Diff}_0(S)$!

In fact this is not as bad as it may seem. Quite often in the quantization of systems with constraints, the physical states are not really vectors in the kinematical state space, but only vectors in some larger topological vector space having the kinematical state space as a dense subspace. Consider a simple example: the particle on the line, regarded as a particle on the plane whose position (q_1, q_2) is subject to the constraint $q_2 = 0$. Naively, one would start with the kinematical state space $L^2(\mathbb{R}^2)$ and look for states ψ satisfying the constraint $\hat{q}_2\psi = 0$, i.e.,

$$q_2\psi(q_1, q_2) = 0.$$

The only L^2 function on the plane satisfying this equation is zero! However, there are *distributions* on the plane satisfying this equation. The space of such solutions is not itself a Hilbert space. However, it has a subspace isomorphic to $L^2(\mathbb{R})$, given by the distributions of the form $\psi(q_1)\delta(q_2)$. This subspace is correct Hilbert space for the particle on the line.

Similarly, while there are practically no solutions of the diffeomorphism constraint living in $L^2(\mathcal{A})$, there are plenty of 'distributional' ones. In fact, there are plenty of generalized measures on $\mathcal{A}$ invariant under small diffeomorphisms of S. The most interesting of these are the gauge-invariant ones, since we would like to solve the Gauss law constraint as well. Note that the the inner product on $L^2(\mathcal{A}/\mathcal{G})$ sets up a chain of inclusions

$$\mathrm{Fun}(\mathcal{A}/\mathcal{G}) \subset L^2(\mathcal{A}/\mathcal{G}) \subset \mathrm{Fun}(\mathcal{A}/\mathcal{G})^*,$$

where $\mathrm{Fun}(\mathcal{A}/\mathcal{G})$ is the gauge-invariant subspace of $\mathrm{Fun}(\mathcal{A})$, and its Banach space dual $\mathrm{Fun}(\mathcal{A}/\mathcal{G})^*$ may be identified with the space of gauge-invariant generalized measures on $\mathcal{A}$. The group $\mathrm{Diff}_0(S)$ acts in a consistent way on all these spaces. A natural candidate for a space of simultaneous solutions of the Gauss law and diffeomorphism constraints is the space

$$\mathrm{Fun}(\mathcal{A}/\mathcal{G})^*_{inv} = \{\mu \in \mathrm{Fun}(\mathcal{A}/\mathcal{G})^* \colon \ \forall g \in \mathrm{Diff}_0(S) \ g\mu = \mu\}.$$

This space is not itself a Hilbert space, but it may have some subspace deserving to be called the 'Hilbert space of diffeomorphism-invariant states'. On the other hand, perhaps there are physically relevant diffeomorphism-invariant states that are not contained in $\mathrm{Fun}(\mathcal{A}/\mathcal{G})^*$, but only in some still larger space containing $L^2(\mathcal{A}/\mathcal{G})$. An example would be the 'Chern-Simons state'. To understand these would require further study of generalized functions on $\mathcal{A}/\mathcal{G}$.

Anyway, at least a general recipe for finding elements of $\mathrm{Fun}(\mathcal{A}/\mathcal{G})^*_{inv}$ is known, as are many interesting examples [7, 9]. Among the most interesting are the 'knot states', which appeared already in a nonrigorous way in the pioneering work of Rovelli and Smolin [45]. These are most easily described using spin networks, although they were not originally constructed that way [4]. Fix an isotopy class C of analytic knots and an irreducible unitary representation ρ of SU(2). This determines a set S of spin network states, namely the Wilson loops $\mathrm{tr}(\rho(T\exp\int_\alpha A))$ with $\alpha \in C$. Since linear combinations of spin network states are dense in $\mathrm{Fun}(\mathcal{A}/\mathcal{G})$, any generalized measure $\mu \in \mathrm{Fun}(\mathcal{A}/\mathcal{G})^*$ is determined by its values on spin network states. We define the knot state μ by setting $\mu(\psi) = 1$ if $\psi \in S$ and $\mu(\psi) = 0$ if ψ is a spin network state not in S. One must check, of course, that μ is a well-defined generalized measure, which involves proving a certain bound. By construction μ is invariant under small diffeomorphisms, so we have $\mu \in \mathrm{Fun}(\mathcal{A}/\mathcal{G})^*_{inv}$.

Since a Wilson loop is just a special sort of spin network, it is natural to ask if this procedure generalizes to yield 'diffeomorphism-invariant spin network states'. The idea would be to use an arbitrary isotopy class of spin networks to obtain a set S of spin network states, and to define $\mu \in \mathrm{Fun}(\mathcal{A}/\mathcal{G})^*_{inv}$ by setting $\mu(\psi) = 1$ for $\psi \in S$ and $\mu(\psi) = 0$ if ψ is a spin network state not in S. The problem is simply to prove that μ is a well-defined generalized measure. This seems to be true when we start with an ambient isotopy class of *links* in S labelled with representations, but the general spin network case is more subtle and still under investigation.

Now let us say a bit about the Hamiltonian constraint. This is perhaps the most controversial aspect of the whole subject, and certainly one of the most important ones: if we find a way to rigorously treat the Hamiltonian constraint, we will be quite close to a rigorous quantization of Einstein's equation, but if not, the whole approach described here may be fundamentally flawed, or at least in need of very new ideas. A lot of work has been done on the Hamiltonian constraint, which we cannot really do justice to here [15, 16, 17, 20, 30, 34]. Naively, the problem is to make sense of

$$\hat{\mathcal{H}}(x) = \hbar^2 \epsilon_{abc} F^c_{ij}(x) \frac{\partial}{\partial A^{ia}(x)} \frac{\partial}{\partial A^{jb}(x)}$$

as a distribution on S taking values in operators on some space of holomorphic functions on $\mathcal{A}_+$. Eventually, however, we want to find states that are annihilated by this constraint together with the Gauss law and diffeomorphism constraints. Thus we would be perfectly happy if we could first use the coherent state transform to transfer the constraint to $L^2(\mathcal{A})$, and then find some subspace of $\mathrm{Fun}(\mathcal{A}/\mathcal{G})^*_{inv}$ — or some related space of solutions of the Gauss law and diffeomorphism constraints — that we could argue was annihilated by the Hamiltonian constraint. The problem is that, in constrast to the Gauss law and diffeomorphism constraints, there is no easy geometrical interpretation of the Hamiltonian constraint in terms of connections on S to fall back upon.

At first it might seem foolish to even *hope* for a simple geometrical 3-dimensional interpretation of the Hamiltonian constraint. After all, the Hamiltonian constraint expresses the *4-dimensional* diffeomorphism invariance of general relativity; or in other words, it encodes the dynamics of the theory. It is all the more tantalizing, therefore, that there are some hints of such an interpretation. Rovelli and Smolin's discovery of these was one of the early successes of the loop representation [45], but making them precise has proved to be very difficult.

Stripped of all nuance, the observation of Rovelli and Smolin reduces to the following. Given a knot α in S, the Wilson loop

$$\psi(A) = \mathrm{tr}(\rho(T\exp\int_\alpha A))$$

is a holomorphic function on $\mathcal{A}_+$. Heuristically speaking, when one applies the functional derivative $\partial/\partial A_i^a(x)$ to ψ one brings down a factor of the tangent vector $\dot\alpha^i(t)$ if $\alpha(t) = x$. So the double functional derivative in $\hat{\mathcal{H}}(x)$ brings down a factor of $\dot\alpha^i(t)\dot\alpha^j(t)$, which is symmetric in i and j. Since F_{ij}^c is antisymmetric in i and j, one obtains

$$\hat{\mathcal{H}}(x)\psi = 0.$$

In fact, when one goes through the argument more carefully one discovers that the double functional derivative of a Wilson loop is very singular, so the result $\hat{\mathcal{H}}(x)\psi = 0$ is a purely formal one unless one can devise some regularization procedure to make the argument rigorous. This has not yet been achieved. Still, upon examination, the argument seems to suggest a 3-dimensional geometrical interpretation of the Hamiltonian constraint as some kind of 'shift operator' generating the motion of a Wilson loop, or more generally the edges of a spin network, along its tangent vectors [45]. For this reason, much effort has been expended to understand the argument and find a context in which it can be made rigorous. One would be very happy, for example, to find by some heuristic argument a general formula for the action of the Hamiltonian constraint on spin network states, or perhaps 'diffeomorphism-invariant spin network states', which one could then subsequently justify by means of its good properties.

In short, quantum gravity continues to provide mathematical physics with challenging — indeed quite frustrating — problems. However, let us conclude on a more upbeat note! Even at the level of the kinematical state space $L^2(\mathcal{A}/\mathcal{G})$, there are some very intriguing applications of spin networks to physics. Classically, in the metric representation a state of gravity is described by the metric on S and its first time derivative. We can rewrite interesting functions of the metric in terms of the new variables, and then attempt to quantize them by replacing A and $\tilde{E}$ by their quantum versions in these expressions, obtaining operators on $L^2(\mathcal{A})$. These operators should commute with the action of $\mathcal{G}$, hence give rise to operators on $L^2(\mathcal{A}/\mathcal{G})$. As the example of the Hamiltonian constraint shows, carrying this out is by no means straightforward in all cases. However, Rovelli and Smolin [46] have considered the *area of a surface* and the *volume of a region* in S, which are technically simpler, and obtained explicit formulas for their quantum versions as operators on $L^2(\mathcal{A}/\mathcal{G})$, in terms of the spin network basis. These operators turn out to have discrete spectrum: certain multiples of Planck length squared for the area operator, and certain multiples of the Planck length cubed for the volume operator. In quantum theory, the spectrum of a self-adjoint operator corresponds to the values the corresponding observable can assume, so this is an indication that area and volume are 'quantized' in the very literal sense of assuming a discrete set of values! Moreover, there is a simple geometrical reason for this fact. To speak in rather oversimplified terms, the area operator applied to a given spin network state counts the number of points at which an *edge* of the spin network intersects the surface in question, weighted by a factor of $\sqrt{j(j+1)}\ell_p^2$, where j is the spin labelling that edge. Similarly, the volume operator applied to a spin network state counts the number of *vertices* of the spin network contained in the region in question, weighted by some function of their labellings by intertwiners and the geometry of the incident edges.

Until we have a fully working theory of quantum gravity, and understand how to take the other forces in account, it is dangerous to take too seriously any physical predictions a partial theory might make. Moreover, this prediction of discreteness of area and volume at the Planck scale is absurdly hard to test with present technology! Nonetheless, the idea that the marriage of Einstein's equation and quantum theory could make such a fascinating prediction should serve as a kind of inspiration for mathematicians and physicists working on quantum gravity.

References

[1] A. Ashtekar, New Hamiltonian formulation of general relativity, *Phys. Rev.* **D36** 1587-1602.

[2] A. Ashtekar and invited contributors, *New Perspectives in Canonical Gravity*, Bibliopolis, Napoli, Italy, 1988. (Available through the American Institute of Physics; errata available from the Center for Gravitational Physics and Geometry at Pennsylvania State University.)

[3] A. Ashtekar, *Lectures on Non-perturbative Canonical Quantum Gravity*, Singapore, World Scientific, 1991.

[4] A. Ashtekar, Mathematical problems of non-perturbative quantum general relativity, in *Proceedings of the 1992 Les Houches Summer School on Gravitation and Quantization*, ed. B. Julia, North-Holland, Amsterdam, 1993.

[5] A. Ashtekar and C. Isham, Representations of the holonomy algebra of gravity and non-abelian gauge theories, *Class. Quan. Grav.* **9** (1992), 1069-1100.

[6] A. Ashtekar and J. Lewandowski, Representation theory of analytic holonomy C*-algebras, in *Knots and Quantum Gravity*, ed. J. Baez, Oxford, Oxford U. Press, 1994, pp. 21-61.

[7] A. Ashtekar and J. Lewandowski, Projective techniques and functional integration for gauge theories, to appear in *Jour. Math. Phys.*.

[8] A. Ashtekar, J. Lewandowski, D. Marolf, J. Mourão and T. Thiemann, Coherent state transforms for spaces of connections, to appear in *Jour. Funct. Analysis.*

[9] J. Baez, Diffeomorphism-invariant generalized measures on the space of connections modulo gauge transformations, in *Proceedings of the Conference on Quantum Topology*, ed. D. Yetter, World Scientific, Singapore, 1994.

[10] J. Baez, Generalized measures in gauge theory, *Lett. Math. Phys.* **31** (1994), 213-223.

[11] J. Baez, editor, *Knots and Quantum Gravity*, Oxford U. Press, Oxford, 1994.

[12] J. Baez, Knots and quantum gravity: progress and prospects, to appear in the proceedings of the Seventh Marcel Grossman Meeting on General Relativity, University of California at Riverside preprint available as gr-qc/9410018.

[13] J. Baez, Spin networks in gauge theory, to appear in *Adv. Math.* .

[14] J. Baez and J. P. Muniain, *Gauge Fields, Knots and Gravity*, World Scientific, Singapore, 1994. (Errata available from the authors.)

[15] C. Di Bartolo, R. Gambini, J. Griego and J. Pullin, The space of states in quantum gravity in terms of loops and extended loops: some remarks, Pennsylvania State University preprint CGPG-95/3-5, available as gr-qc/9503059.

[16] M. Blencowe, The Hamiltonian constraint in quantum gravity, *Nucl. Phys.* **B341** (1990), 213-251.

[17] R. Borissov, Regularization of the Hamiltonian constraint and closure of the constraint algebra, Temple University preprint TU-94-11, available as gr-qc/9411038.

[18] B. Brügmann, Loop representations, Max-Planck-Institute preprint MPI-Ph/93-94, available as gr-qc/9312001.

[19] B. Brügmann, personal communication.

[20] B. Brügmann and J. Pullin, On the constraints of quantum gravity in the loop representation, *Nucl. Phys.* **B390** (1993), 399-438.

[21] L. Chang and C. Soo, BRST cohomology and invariants of four-dimensional gravity in Ashtekar's variables, *Phys. Rev.* **D46** (1992), 4257-4262.

[22] L. Crane, L. Kauffman and D. Yetter, State-sum invariants of 4-manifolds, I, Kansas State University preprint, available as hep-th/9409167.

[23] B. DeWitt, Quantum theory of gravity, I-III, *Phys. Rev.* **160** (1967) 1113-1148, **162** (1967) 1195-1239, 1239-1256.

[24] P. A. M. Dirac, *Lectures on Quantum Mechanics*, Yeshiva University, New York, 1964.

[25] S. Donaldson and P. Kronheimer, *The Geometry of Four-Manifolds*, Oxford U. Press, Oxford, 1990.

[26] B. Durhuus, H. Jakobsen and R. Nest, Topological quantum field theories from generalized $6j$-symbols, *Rev. Math. Phys.* **5** (1993), 1-67.

[27] A. Fischer and J. Marsden, The initial value problem and the dynamical formulation of general relativity, in *General Relativity, an Einstein Centenary Survey*, eds. S. Hawking and W. Israel, Cambridge U. Press, Cambridge, 1979, pp. 138-211.

[28] T. Foxon, Spin networks, Turaev-Viro theory and the loop representation, *Class. Quan. Grav.* **12** (1995), 951-964.

[29] D. Freed and K. Uhlenbeck, *Instantons and Four-Manifolds*, Springer-Verlag, Berlin, 1984.

[30] R. Gambini, Loop space representation of quantum general relativity and the group of loops, *Phys. Lett.* **B255** (1991), 180-188.

[31] R. Gambini and A. Trias, Gauge dynamics in the C-representation, *Nucl. Phys.* **B278** (1986), 436-448.

[32] B. Hall, The Segal-Bargmann coherent state transform for compact Lie groups, *Jour. Funct. Analysis* **122** (1994), 103-151.

[33] C. Isham, Canonical quantum gravity and the problem of time, in *Integral Systems, Quantum Groups, and Quantum Field Theories*, eds. L. A. Ibort and M. A. Rodriguez, Kluwer, Dordrecht, 1993, pp. 157-207.

[34] T. Jacobson and L. Smolin, Nonperturbative quantum geometries, *Nucl. Phys.* **B299** (1988) 295-345.

[35] L. Kauffman, *Knots and Physics*, World Scientific, Singapore, 1991.

[36] L. Kauffman and S. Lins, *Temperley-Lieb Recoupling Theory and Invariants of 3-Manifolds*, Princeton U. Press, Princeton, 1994.

[37] J. Lewandowski, Topological measure and graph-differential geometry on the quotient space of connections, International Journal of Theoretical Physics, **3** (1994) 207-211.

[38] P. Peldan, Actions for gravity, with generalizations: a review, *Class. Quan. Grav.* **11** (1994), 1087-1132.

[39] R. Penrose, Angular momentum; an approach to combinatorial space time, in *Quantum Theory and Beyond*, ed. T. Bastin, Cambridge University Press, Cambridge, 1971.

[40] G. Ponzano and T. Regge, Semiclassical limits of Racah coefficients, in *Spectroscopic and Group Theoretical Methods in Physics*, ed. F. Bloch, Wiley, New York, 1968.

[41] J. Pullin, Knot theory and quantum gravity: a primer, in *Proceedings ofthe Vth Mexican School of Particles and Fields,* ed. J. Lucio, World Scientific, Singapore, 1993.

[42] N. Reshetikhin and V. Turaev, Ribbon graphs and their invariants derived from quantum groups, *Comm. Math. Phys.* **127** (1990), 1-26.
Invariants of 3-manifolds via link polynomials and quantum groups, *Invent. Math.* **103** (1991), 547-597.

[43] J. Roberts, Skein theory and Turaev-Viro invariants, to appear in *Topology.*

[44] C. Rovelli, The basis of the Ponzano-Regge-Turaev-Viro-Ooguri model is the loop representation basis, *Phys. Rev.* **D48** (1993), 2702-2707.

[45] C. Rovelli and L. Smolin, Loop representation for quantum general relativity, *Nucl. Phys.* **B331** (1990), 80-152.

[46] C. Rovelli and L. Smolin, Discreteness of area and volume in quantum gravity, University of Pittsburgh preprint available as gr-qc/9411005.
C. Rovelli and L. Smolin, Spin networks in quantum gravity, University of Pittsburgh preprint.

[47] V. Turaev and O. Viro, State sum invariants of 3-manifolds and quantum $6j$ symbols, *Topology* **31** (1992), 865-902.

[48] E. Witten, Quantum field theory and the Jones polynomial, *Comm. Math. Phys.* **121** (1989), 351-399.

[49] E. Witten, 2+1 dimensional gravity as an exactly soluble system, *Nucl. Phys.* **B311** (1988) 46-78.

DEPARTMENT OF MATHEMATICS, UNIVERSITY OF CALIFORNIA, RIVERSIDE CA 92521,

E-mail address: baez@math.ucr.edu

Index

Other Titles in This Series

(Continued from the front of this publication)

ISBN 0-8218-0380-8

9 780821 803806